WHAT PRICE VIGILANCE?

The Burdens of National Defense

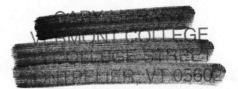

WHAT
PRICE
VIGILANCE?

THE BURDENS OF NATIONAL DEFENSE

by Bruce M. Russett

New Haven and London, Yale University Press, 1970

Library of Congress catalog card number: 75-119475
ISBN: 0-300-01358-2 (cloth), 0-300-01359-0 (paper)
Designed by Sally Sullivan,
set in IBM Selectric Press Roman type,
and printed in the United States of America by
The Carl Purington Rollins Printing-Office
of the Yale University Press.
Distributed in Great Britain, Europe, and Africa by
Yale University Press, Ltd., London; in Canada by
McGill-Queen's University Press, Montreal; in Mexico
by Centro Interamericano de Libros Académicos,
Mexico City; in Australasia by Australia and New
Zealand Book Co., Pty., Ltd., Artarmon, New South
Wales; in India by UBS Publishers' Distributors Pvt.,
Ltd., Delhi; in Japan by John Weatherhill, Inc.,
Tokyo.

To Cynthia
sine quo non

Contents

Preface

I wrote this book because I became concerned for my country, about the effect on American society of continued high levels of military spending. In this sense it stems much more directly than have my earlier books from an immediate interest in public policy. At the same time, I wanted the work to be as much as possible a product of modern social science theory and methods for dealing with evidence. The insistence on following the rules of contemporary methodology is especially important in this substantive area, in which until now the overwhelming majority of studies, however original and interesting, have relied largely upon intuition and anecdote as a basis for argument. Just because the whole question of a "military-industrial complex" is so controversial, it is essential, if we are to understand the problem of defense spending and eventually to direct our resources into more productive channels, to be as careful as we can in examining the evidence. It is also important, whenever possible, to make general statements about political relationships rather than simply descriptive statements narrowly bound to particular phenomena of the past. In this sense one ought to contribute to the growth of social science rather than merely use it, in the spirit of engineering, to analyze and solve some special immediate problem.

I cannot pretend to have succeeded on all these counts. Most importantly, the book is much more a work of social engineering than of science. Save perhaps for some of the sections on arms races and on burden-sharing in alliances, the findings are more time- and space-bound than I would like. Yet that was inherent in the decision to write, above all, a policy-relevant book, and I make no apologies for it. Furthermore, the book is ultimately a combination of "hard" social science with the more traditional tools of analysis. As we all recognize, some questions are more susceptible to quantitative treatment than are others, and the whole problem of the causes and consequences of military spending is in reality a congeries of questions, some rather readily subject to hard analysis, others quite intractable. Thus the work includes several quantitative sections, bridged and extended by inference and, ultimately, by speculation. I have, nevertheless, tried to make clear to the reader what conclusions I felt were firmly based and what others both he and I ought to weigh carefully. Finally, because the book was inspired by a policy concern, I did not feel I should hold back on publication until I had completed all imaginable analyses. Given time, for instance, I would have liked to build the conclusions about congressional behavior on military matters upon an examination of voting in both houses over a time span of an entire decade or longer. But with the intense political debate on these matters and the impact of the decisions being taken, I could not justify long delay. I did not compromise on the rigor of the analyses I did undertake. I regard this book as a direct rebuttal to the charge that quantitative social science has nothing to say about the really important issues of American politics.

Because I insisted on using quantitative analysis wherever possible, some portions of the book may be hard reading for those who are not accustomed to such methods but come to the book because of its policy focus. I have tried to ease that difficulty by simplifying the methodological discussion as much as possible and relegating much of it to footnotes and an appendix.

I am indebted to many individuals and organizations for help with the book. Among the latter, research grants from the National Science Foundation (No. GS 614 and 2365) and a contract from ARPA, Behavioral Sciences (Contract No. N-0014-67-A-0097-0007, monitored by the Office of Naval Research) were indispensable. In view of the controversial nature of the book's concerns, however, it is essential to be clear about what was and what was not supported by public funds. NSF and ARPA funds were used, in connection with a broader program of studies under my grants and contracts, for the critical examination of the arms race literature in the appendix. ARPA monies were also used for the analysis of alliance burden-sharing in chapter 4 and the comparative study that constitutes chapter 6. Also, my funds from NSF contributed to chapter 5 on American resource allocation and to the methodological work by John McCarthy on the scaling techniques employed in chapter 2. No public agency, however, is responsible for the substantive conclusions of chapter 2 or of chapter 3, for the focus of the book or the general conclusions of the introductory and closing chapters, or indeed for my interpretations anywhere. The substantive work on senatorial voting and its correlates was supported entirely by Yale University with institutional funds. Without that private research money the book would not have been possible. I wrote up almost all of the material as Visiting Professor at the Institut d'Etudes Européennes of the University of Brussels and while on a fellowship from the John Simon Guggenheim Foundation. It must be quite clear that no individual or organization, public or private, is responsible for errors of fact or interpretation.

There are two exceptions to the last absolution. The

work in the appendix, on mathematical models of arms races, is entirely the work of Peter A. Busch, done especially for this book. It fills a critical area in any examination of the causes of military expenditures, and I am most happy to be able to include it. In addition, my student Harvey Starr shares with me the authorship and the responsibility of chapter 4 on alliances. Others to whom I am indebted but who bear no responsibility include John L. McCarthy of the Yale History Department, whose development and application of scaling analysis was indispensable, and my hard-working and dependable research assistant, Wayne Moyer. In an undergraduate paper Thomas Weil did some preliminary analysis that contributed to chapters 2 and 3. Caroline Stancliff typed parts of the manuscript and performed some of the computations; Patricia Stannard typed most of it, often from handwriting that would have been unintelligible to anyone else. John Stannard drew the graphs. It was again a pleasure to work with Marian Neal Ash as editor. My presence in Europe during production made her responsibilities heavier than usual.

Portions of chapters 5 and 6 were published previously in the *American Political Science Review* and are reprinted with permission. These chapters, however, have been revised and very greatly expanded with new material.

B. M. R.

Overijse, Belgium
January 1970

1. Climbing High Into the Sun?

The Growing Burdens of Defense

This is a book about a central issue of American politics—expenditures for national defense. It is about their current high level—why they are high and what some of the consequences are for American politics, the economy, and the society. While not a complete study of these questions, it brings to light information not previously available and takes a look at them from some new perspectives. It assumes that a failure both to answer them and to act on the basis of the answers would be fatal to the kind of nation that the founders meant to bequeath to us and that most of us cherish. But though the exploration is grounded in that passion, it demands objectivity in analysis. We can react too violently as well as too passively against the offenses of defense, or we can from ignorance react against small faults while missing great ones. Or in shunning the violence of others' attacks, we can close our eyes to abuses we should not tolerate. So within the limits of the human weakness of author and reader, let us weigh carefully some evidence about why Americans must look within their own borders for much of the impetus to military spending and what its effects on the country have been.

Large armed forces have become a part of American life, as American as cherry pie. No American under fifty can remember a period in his adult lifetime when our armed forces were small. Although it has but the fourth largest population among the world's peoples, our country maintains the largest armed forces on earth. In 1968 we had 3,500,000 men under arms, more than Russia (3,220,000) or China (2,761,000).[1] While this did not

quite represent the highest *proportion* of able manpower under arms of any nation, even in 1965 before
the Vietnam buildup it was fairly close (17th out of
121).[2]

Yet a large army is not a part of this nation's heritage.
George Washington was a general, but he neither led nor
left a standing army; his responsibility for a large military establishment is about as certain as his responsibility for having cut down the fabled cherry tree. During
the first 170 years of our history, only in actual wartime
or in the year or two immediately afterward did our
armed forces ever employ as many as 1 percent of the
working-age male population. It is since World War II
that the nation's living style has changed. We have at no
point since 1941 had fewer than 1,400,000 men under
arms nor as little as 3 percent of working-age males in
the armed forces; since the Korean War twenty years
ago that figure has not fallen below 5 percent. The same
pattern appears when we look at military expenditures.
Before 1939 the peacetime military budget was barely
higher than 1 percent of the gross national product. The
post-World War II floor was 3.9 percent in 1947 and
since Korea has fluctuated between 7.3 and 11.3 percent.

Obviously, Americans have come to tolerate a peacetime level of military spending and manpower unprece-
dented in their history. Why did such high levels come
about, and what have their consequences been for
American life? There are peculiarities about the growth
of the American military that suggest partly competing
and partly complementary hypotheses about its development.

Most important of the peculiarities is what we may
call the "ratchet effect" of wars. According to a report

of the National Industrial Conference Board, the proportion of the American budget devoted to military expenditures has never, after any war, returned to the prewar level.[3] Figure 1.1 shows military personnel as a

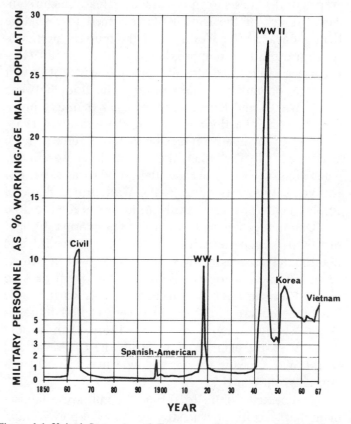

Figure 1.1 United States Armed Forces as a Percentage of Male Labor Force, 1850-1967
Sources: U.S. Bureau of the Census, *Historical Statistics of the United States, Colonial Times to 1957* (Washington: U.S. Government Printing Office, 1960) and U.S. Bureau of the Census, *Statistical Abstract of the United States, 1962* (Washington: U.S. Government Printing Office, 1962); also *Statistical Abstract, 1963-68.*

percentage of working-age males since 1850. Working-age males form the denominator to gauge the productive loss to the labor force represented by military service.

Beginning with the Spanish-American War in 1898, every conflict has shown a marked increase in the standard "floor" of the United States armed forces. More than a decade after World War I the military participation ratio did fall to approximately the level of 1916—when war clouds had already appeared—but not to that of 1915 or earlier in the century. The floor between World War II and Korea was far higher than any previous one, and the floor was again elevated after 1953, never again dropping appreciably below 5 percent. Even after the Civil War, when the prewar level was finally reached again after a decade, the decline was slow. For ten years Indian fighting in the West provided useful employment for martial skills no longer needed to hold the Union together or to fight foreign nations. In light of this striking pattern one wonders how successful, and if ultimately successful how quickly so, will be efforts to shrink the American military establishment back to its pre-Vietnam level. And in passing, it is worth noting how the modern, fully industrialized United States is able to support a much larger armed force than it could do as a partly agrarian state. Both the Civil War and World War I peaks are far below what was achieved during World War II.[4]

The repeated failure to shrink the military establishment back to its prewar strength shows up even more clearly in the data on absolute numbers of military personnel. They indicate a virtual doubling of the armed forces after every war. From 1871 to 1898 the American Armed forces numbered fewer than 50,000; after the Spanish-American War they never again dropped below 100,000. The aftermath of World War I saw a leveling-off to about 250,000, but the World War II mobilization left 1,400,000 as the apparent permanent floor. Since the Korean War the United States military establishment has never numbered fewer than about 2,500,000 men.[5]

The Superpower Hypothesis

Surely the most obvious explanation is that ever higher military expenditures have been forced upon this nation by the dynamics of international conflict and world leadership. During the nineteenth century Americans could afford to dispense with a standing army or large navy. Their country was protected, by distance and the mighty oceans, from the centers of European power. Of our neighbors, not Canada, nor Mexico, nor the western Indians posed a threat to national survival. On the occasions when Europeans stopped fighting each other long enough to contemplate expansion in the western hemisphere, the British fleet stood ready to discourage them. In turn, Canada was always potentially vulnerable to an American attack, so Britain was not tempted to use its seapower against our almost undefended coasts. By the turn of the century, however, our isolation became less secure. Improved naval technology and then aircraft reduced the oceans' protection; the world wars threatened to destroy the European balance of power that had kept any single state from dominance of the old world. In 1945 that balance was utterly shattered, and to preserve its security the United States was forced to maintain a standing army and the world's most powerful navy. Thus the gradual growth of our military establishment could be explained by gradual changes in the world environment and this country's reluctant accession to superpower status and global leadership. The army stayed larger after each war because objectively there was more for it to do.

Furthermore, the late 1940s and most of the 1950s were characterized by what seemed a very clear and present danger to American security. Whatever current social scientists or revisionist historians may now declare about the reality of that threat in the early cold war years, few men said at the time that the danger was not real. Henry Wallace's ringing defeat in the 1948 presidential election, followed by the rise of Joseph McCarthy and the political climate that nurtured him, marked an end for over a decade to public questioning of the cold war's necessity. The army remained large after the Korean war because of what seemed a grave and obvious menace from the communist bloc.

It is not possible to refute the argument about new American responsibilities in the twentieth century and a new foreign threat; on the contrary, it is plausible and surely in large part correct. Still, it is incomplete. Why was there such a discrete jump in the military "floor" immediately after the splendid little war of 1898? Surely any changes in the external environment then were gradual, not themselves requiring a sudden doubling of the armed forces. Secondly, the pattern of military activity after World War I does not quite fit this explanation. Surely 1917 should have provided a vivid demonstration of the loss of American security through isolation. The development of long-range aircraft in the twenties and thirties should, by this logic, have produced a steady growth in our armed forces. But instead they ultimately slipped back—after a decade—to the 1916 proportionate manpower level. That fact, coupled with the ten-year delay, suggests a certain "stickiness" in military levels that must be accounted for by other factors. Finally, the picture of the cold war as America's

response to an external challenge is surely an oversimpli-
fied one. Our understanding of the dynamics of arms
races now leads us to see it as a period at least in part of
mutual challenge and response between the United
States and Russia. Each has been in some measure responding to the other's military expenditures; while
America has reacted to the Russian stimulus, it has also,
perhaps unavoidably, provoked Russian countermeasures that were in turn seen as demanding new
American purchases in the spiral.

The Arms Race Hypothesis

Thus the arms race pattern of interaction between two
conflicting powers is widely regarded as the correct explanation of recent United States arms spending. Combined with the simpler image of necessary response to
Soviet aggression, it is tempting to look no farther. But
to stop with these assertions, however plausible, would
be scientifically unsatisfying. Furthermore, one would
like to know (a) the relative importance of the arms race
vs. the pure American response to aggression, and (b) to
the degree it does apply, exactly what the characteristics
of the arms race pattern are. If we hope someday to
control the spiral we must know just what propels it, in
more detail than can be provided by the simple verbal
explanation. Hence there have been many recent efforts,
stimulated by the early work of Lewis Frye Richardson,
to develop rigorous mathematical models of arms races
that might be applied to the experience of the last two
and one-half decades.
  Many of these models are evaluated in detail in an
appendix to this book, written by Peter Busch. Some
are purely deductive models; others test the deductive
system with data from various arms races of the last

century, including the current competition between the
United States and the Soviet Union. It is unmistakably
clear from this review that none of the models as currently constituted can adequately explain the Soviet-
American situation. Several reasons are elaborated for
this failure:

1. Most obvious, but not necessarily most important,
is the unreliability of the published data on military
expenditures for many countries. This is true in some
degree for analyses of the pre-World War I and pre-
World War II years, but most acutely for analyses of the
recent behavior of the Soviet Union and China. In these
cases it is difficult even for an expert on the country to
know whether apparent trends reflect actual variations
in military spending or merely changing fashions in data
manufacturing.

2. The quantity of data is another limitation, one related to the first. Frederic L. Pryor, for example, has
put together quite good data on the Soviet Union, but
only for 1950-62.[6] Any attempt to analyze arms races
must at least consider the possibility that the relevant
influences and perhaps the basic relationships differ
with historical periods. Clearly the major powers and
the relationships among them changed greatly during
World War I and again during World War II. For the
Soviet-American situation one would want to allow for
the possibility of major change with the death of Stalin
or the election of President Kennedy. But this means
that the number of observations (usually fiscal years)
for any one period is too small to permit tests sufficiently complex to be interesting. Often the analyst tries to
reduce this problem by bringing in new data to estimate
some of the influences. But unfortunately so many in-
valid assumptions are made that the value arrived at is

hardly more than sheer guesswork.[7] By using doubtful outside information and very dubious assumptions, complex relationships are seemingly tested, when in fact a much simpler relationship is applied to the data. The result is that a study often seems to show that the data support a particular model, whereas it actually indicates only that *some* relationship fits the observations.

3. Because data are poor and few, the most important criterion of an arms race model must be the plausibility of its initial assumptions. All the components must have some clearly specified meaning for political behavior—and this is true not only for the parts of a system but for the system as a whole. As is argued in the appendix, too many of the arms race models fail to meet this test.[8]

I certainly do not mean to condemn all attempts to build rigorous models of arms race phenomena. They get us away from a perspective of blaming all our troubles on our opponent's aggressiveness. Any understanding of contemporary arms purchases will have to consider the mutual perceptions of the antagonists and their interactions, an adequate explanation will have to use mathematics, and it may be that models derived from such other fields as physics will provide valuable insights. Several extant arms race models do indeed seem potentially promising. But the fact remains that for a combination of theoretical and empirical reasons no such model has yet been applied satisfactorily to the Soviet-American rivalry. Exclusive concentration of research efforts on arms race models risks bogging down the enterprise in one of the most intractable social science problems. Furthermore, a focus on international conflict neglects some key phenomena. Without ignoring the importance of interaction between the super-

powers, an examination of recent great power military
expenditures therefore must also proceed along other
lines.

Two considerations support this procedure. First, in
the work previously cited Pryor correlated changes in
American defense expenditures with those of the Soviet
Union for each year over the 1950 to 1962 period.[9]
Nearly half the variation (48 percent) in one nation's
expenditures could be explained by the other's expenditures. Despite the short time period involved, this finding is statistically significant and provides good evidence
that the two states either are interacting or, at the least,
that one is responding very quickly to changes in the
other's expenditure. But at the same time it is still less
than half the variation that is being explained. Part of
the problem is doubtless one of data quality, yet one is
compelled to investigate further. International influences other than those of direct competition, and domestic influences arising out of the political systems of
one or both powers, must be considered.

Secondly, all of the arms race models we have found
do themselves incorporate domestic influences. Critical
components typically take into account "aggressiveness," intelligence capabilities, "grievance" terms, and
"fatigue." These are frequently assumed to be different
for different powers and at different times. Thus one
cannot even employ arms race explanations without detailed attention to a state's relations with its allies—its
alliance system—and to internal factors.

Alliances

Alliance relationships are important because our allies
are also in potential conflict with the Soviet Union and

also should feel the pressures of an East-West arms race. A major American goal in organizing NATO and other alliances was to pool resources, so that defense would cost each member less than if all went their separate ways. American leaders reasoned that by adding Western European ground forces to its strength, for example, the United States could get by with a smaller army of its own.

Yet despite its many alliances, on a list of defense expenditures as a percentage of GNP for all the 121 nations of the world in 1965 America ranked 15th, surpassed only by beleaguered Albania, Russia, and a few nations involved in local situations of sporadic violence (Israel, a few Arab countries, Laos, the two Vietnams, the two Chinas, and Indonesia).[10] Even West Germany, which one might think of as most directly threatened by the communists, spent an amount that was, for its size, less than two-thirds of America's.

America's intention to ease its own military burdens was clearly in conflict with another goal pursued simultaneously, namely to shelter its allies, to bring them under the protection of this nation's deterrent umbrella. To the degree that Germany, Japan, and Brazil, let us say, are confident of American protection, they need not build up their own military forces but may turn their resources to economic development or other purposes. This has become a major source of friction, particularly in NATO, and many congressmen demand that if Europeans do not do more to defend themselves, the United States must reduce its commitments.

It is tempting to blame the individual allies for this situation, but the Soviets now find themselves in an analogous position. There are in fact sound theoretical reasons why smaller nations are able to reduce their defense burdens by alliance more than are larger ones, and why the East-West arms race primarily afflicts the superpowers. We shall examine both the theory and relevant data in chapter 4.

Defense and Domestic Politics

Most importantly, we must focus on domestic influences. Chapters 2 and 3 do so in detail, especially on congressional politics, but we can make some general observations here. An example of their importance is the widespread assumption that former President Johnson's 1968 decision to proceed with his Sentinel ABM system was to protect himself from right-wing political critics who were already unhappy about lack of progress in Vietnam—not out of any great sense of need for the system to protect the national security. Former Budget Director Charles Schultz summarized his view of American defense options in a way that strikingly suggests a combination of international constraints and opportunities for choice as permitted by domestic politics:

> In short, the future budgetary consequences of present strategic policy may prove to represent an unstable equilibrium. Either decisions will be made to reduce those expenditures, or they may themselves create a situation in which further expenditure increases will occur. To the extent that this evaluation is correct, the post-Vietnam fiscal dividend will either be significantly increased by policies that reduce military spending, or it will be significantly eroded by further additions to that spending. There may be no intermediate position.[11]

In other words, the United States had a choice, but a

limited one. If the purchase of such systems as ABM and MIRVs were avoided or postponed, then there might be excellent prospects for reducing the defense budget. But if these systems were procured, even in relatively small quantities that initially required no increase in the overall defense budget, they probably would initiate Russian countermeasures that America, in turn, would feel obliged to meet. Thus a spiral of mutual "countermeasures" would be set off to inflate both nations' arms spending far above the level directly implied by the initial choice. Domestic factors therefore may determine whether our interactive spending with the Russians is carried on at high or at low levels; alternatively we may consider that arms races merely establish a range of options, perhaps rather wide, within which domestic influences make the difference.

One kind of domestic explanation, with substantial plausibility, is Parkinson's Law about the virtual impossibility of disbanding any large organization. Thus when continental air defense loses its raison d'être with the diminution of the threat from manned bombers, what could be simpler than to look for a new, related activity such as ABM?[12] Another is the set of various economic explanations for imperialism and war. Such theories are at least as old as capitalism itself, but they date in their most articulated versions from J. A. Hobson and Lenin around the turn of the last century. They still appear, often in sophisticated form, tracing the dynamics of foreign conflict to the pressures of a decaying capitalism. In its most simplified version the proposition is that for capitalist economies in general, and for the United States in particular, the overall "health" of the economy requires high levels of military spending to take up slack and maintain demand in an otherwise depression-prone economy. The Marxist variant is that, impelled outward by the falling rate of interest at home, capitalists seek colonies or quasi-colonies for new investment opportunities abroad; the competition between investors of different countries inevitably involves their governments and provokes both war among the imperialist powers and ultimately colonial wars of national liberation by the exploited.[13]

On the whole, these general arguments are not convincing. It is not too difficult to show that many industrialists were unenthusiastic about Britain's nineteenth-century colonial expansion and opposed the Spanish-American War in the United States. Peace rumors on Vietnam feed the bulls, not the bears, on Wall Street, and the Vietnam War has effectively killed any lingering notion that war is good for the American economy. Despite the apocryphal *Report from Iron Mountain,* it seems clear that with the demonstrable public and private needs of this society, and with modern tools of economic analysis and manipulation, full or near-full employment of resources would be maintained even in the face of major cuts in military spending.[14]

On the other hand, it is not hard to show that particular investors have involved their governments in particular wars, even though the business community in general may have been lukewarm.[15] And individual industries—specifically arms manufacturers—may benefit from war even though the economy as a whole may suffer. For example, American revisionist critics of their country's entry into World War I conducted investigations in the 1930s to demonstrate the guilt of munitions-makers. Senator Nye and others blamed this country's military involvement on the "merchants of death." It is of course one thing to demonstrate the high profits made by munitions-makers from wars and quite another to show that the arms merchants significantly affected the decision to go to war.

The "Military-Industrial Complex"

Modern versions of the arms-makers theme abound; they rarely blame the weapons industry for actual war, but they do credit excessive weapons procurement to a confluence of interests of Pentagon and industry in the "military-industrial complex." In essence, the argument is that the cold war and its arms race have bred a new species of industrial firms that produce military equipment, have grown accustomed to the product requirements of the armed forces, and fear they would find it difficult to convert their production to civilian needs. Similarly, the military depends for its prestige and sustenance on the frequent acquisition of new weapons. This is particularly acute when one thinks of the military not as a monolith but considers competition among the major services and among branches within the services. Thus the Navy in the 1950s sought a strategic war capability, first in aircraft carriers and later in missile-carrying submarines, to avoid being submerged by the Air Force; likewise Army personnel sought an anti-missile defense system, and within the Air Force the advocates of a manned bomber capability have fought to keep the Air Force from being grounded, for strategic war purposes, in unglamorous missile silos deep below the western plains. The complementary interest of weapons producers in selling new generations of arms and of the armed forces in buying the weapons is thus said to produce a symbiosis that is mutually profitable for them but costly to the taxpayer and to federal spending for civilian programs.

This proposition is supported by the obvious waste in many past military procurement efforts that either were aborted after costing billions of dollars or were obsolescent or redundant as soon as the equipment came into use and by the questionable military utility of certain current programs. The ABM and the new manned bomber are of warmly debated value. The proposition is further supported by assertions or analyses about the amount of waste in contemporary military budgets. For instance, the recent *Report of the Congressional Conference on the Military Budget and National Priorities* declares:

> Robert Benson, a former Defense Department official, has suggested that even without any change in our foreign policy or military objectives, the Pentagon could save at least $9 billion by eliminating such programs as the manned orbiting laboratory (which duplicates work already being done by NASA), reducing inefficient turnover in the assignment of military officers and effecting other efficiencies in the use of military manpower, changing contracting practices so that greater performance and economy is required of defense contractors, eliminating unneeded and obsolete weapons systems, and cutting back some of our unnecessary overseas troop deployment.[16]

Similarly, *Congressional Quarterly* reports that "Several highly placed Administration and industry sources said cuts totalling $10 billion to $15 billion could be made in the defense budget while retaining or improving the current level of national security." The Army's large supply and maintenance "tail," the ABM, manned strategic bombers, and anti-submarine warfare systems, especially aircraft carriers of high cost and dubious effectiveness, are cited as prime candidates for the fat-trimmer. Further, according to *Congressional Quarterly* one Pentagon source declared "Any increment over $40 billion in the post-Vietnam defense budget will be purely due to pressures by the military-industrial complex rather than to the defense needs of the nation." A distinguished analyst outlines an annual $50 billion post-Vietnam military budget as being entirely feasible and appropriate. That amount would imply a cut of about $10 billion in non-Vietnam expenses.[17]

Unfortunately there is no general agreement on the amount or constitution of Pentagon waste. In these matters, one man's fat is another man's meat. Secretary Laird insists that further cuts in the defense budget would imperil the nation's security; both the ABM and manned bombers have intelligent defenders who are devoid of direct interest in their procurement. Economist Malcolm W. Hoag insists that it would be hard to cut the defense budget very much after Vietnam's costs are subtracted. He outlines one possible cut, of only about $5 billion, as requiring such force changes as moving three Army divisions from overseas to inactive reserve status, deleting the two newest and most expensive tactical air wings, forgoing a new aircraft carrier, and reducing the present carrier force from 15 to 14.[18] While almost everyone can find examples of what he regards as wasteful procurement, others will contest his assessment. Some "unnecessary" procurement must be expected in any government purchasing program, and while one may strongly suspect that American military expenditures are higher than security requires, it is difficult, in the face of conflicting expert testimony, to prove that the level is up into the wild blue yonder.

Nevertheless, it is clear that the market for military

equipment is very different from the classical market of perfect competition and that a variety of unusual influences affect procurement decisions. The dependence of particular firms on Department of Defense contracts is often extremely striking. Of the 44 largest DoD contractors in 1958, 27 made more than 60 percent of their total sales to DoD.[19] Furthermore, it might be most difficult for many of these firms to sell profitably on the civilian market. Especially in the aerospace and shipbuilding industries, they are accustomed not to mass production of items with a relatively low unit cost but to the manufacture of one or a few units, each very expensive and requiring very high performance and reliability characteristics. A Poseidon missile or submarine is most unlike the typical product of civilian-oriented industry, not only in its purposes but in its mode of production. Cost often is much less important than performance; the deliberate decision not to worry much about cost and to employ very high quality components with built-in redundancy is very unlike the mass production of automobiles, with great cost-consciousness, that made Detroit famous. "The relative standing of a firm in the weapons industry reflects more its ability to adapt to new specialties and to exploit new technical possibilities rather than an ability to produce the established products efficiently."[20] Hence the often essential practice of awarding cost-plus contracts, rather than taking competitive bids for a fixed price. In addition, the firms as a matter of course accept high risks in the development of particular products; their contractual agreements with the government largely compensate them for these risks, but they would find it hard to make similar arrangements on the civilian market.

Peck and Scherer note that in the years preceding

their 1962 study the turnover of the top 100 defense contractors was about twice as great as that of the 100 largest industrial firms, overall. From this they concluded that the individual firms faced a good deal of uncertainty. For a particular product, a firm may have few, if any, competitors. But threatened changes in the product mix, coupled with a common lack of flexibility to make something else for military or civilian use, intensify the pressures to obtain a particular military contract.[21] This is perhaps especially true for the decade of the 1950s, with the shift in procurement from conventional military weapons to the exotica of the missile industry. At the same time, the turnover should not be exaggerated. According to a more recent analysis of the largest firms, 18 of the top 25 DoD contractors in 1958 were still in the top 25 in 1967. Hence the same few companies remain heavily dependent on the continuing flow of military orders.[22]

Back in 1962, a subcommittee headed by then Senator Hubert Humphrey mailed inquiries to a sample of DoD contractors asking whether they were willing to shift production to civilian goods if an arms reduction agreement should necessitate cutbacks in defense purchases. The report was that by and large American defense manufacturers not only could make the adjustment but would be delighted to shift to full peacetime production. The mode of investigation by the subcommittee, however, seems a bit naïve. Two years later a different senate subcommittee, chaired by Senator Joseph Clark, concluded that certain firms, especially in the aerospace industry, suffered from an overcapacity that drove them to insist on more missiles than than the nation needed.[23] Peck and Scherer declared that it was precisely the uncertainties and the need to maintain large research staffs to find and exploit new technologies that led to excess capacity in firms' technical groups. According to them, that excess capacity then brought on "goldplating" of the final product.[24]

If this is the case, then many American weapons-manufacturing firms must be very anxious to maintain the flow of military contracts, despite a booming economy and their hypothetical ability to shift to the production of goods for civilian use. That is where the alleged symbiotic relationship between producer and contractor comes in. It is assisted by careful cultivation of personal contacts and the use, by industry, of representatives not only with close knowledge of military needs but with social and professional ties to Pentagon personnel. According to a tabulation released by Senator William Proxmire in March 1969, the 100 largest defense contractors employed a total of 2,072 retired military officers who, when in service, had held ranks at least equivalent to colonel or Navy captain. This figure compares with 721 such former officers in 1959.[25] The literature on lobbying is full of examples of mutually rewarding relationships between Pentagon personnel and their civilian suppliers. Especially intriguing are reports on how an executive agency and a private firm will coordinate their lobbying so as most effectively to influence Congress and the public. For instance, it is said that Army officials suggested to Western Electric that the firm should place commercial advertisements for its proposed Nike missile system.[26] As a reasonably sympathetic observer puts it,

The professional leaders of the U.S. military services, although restrained by American traditions and handicapped by a number of other political liabilities, have felt themselves increasingly constrained by the nature of the American political system and other factors after the turn of the century to engage ever more actively in bureaucratic political struggles in order to prepare for their assigned missions.[27]

The Navy has perhaps been the most discreet of the services in its political activity, concentrating on developing and maintaining good relations with key members of the legislative and executive branches and preferring to avoid public relations efforts designed to affect decision-makers indirectly. Its cultivation of personal relationships slipped at one point, however, becoming inferior to the job being done by the Air Force and Army: "The Navy had so seriously neglected its fence-mending on Capitol Hill by the mid-1950's that Congress asked it to do a better job."[28] Fortunately for the Navy, it did improve.

Adam Yarmolinsky, a former Deputy Assistant Secretary of Defense, observes:

> Surely the military-industrial-congressional complex is not a conspiracy. But there are coincidences among the military project officer who is looking for a star, the civilian who sees an opening for a new branch chief, the defense contractor who is running out of work, the union business agents who can see layoffs coming, and the congressman who is concerned about the campaign contributions from business and labor as well as about the prosperity of his district. Each of these constellations of interests wants to expand the defense establishment in its own direction.[29]

It is not surprising, nor is it censurable, that the armed services and their suppliers should undertake political activities. Virtually all firms that sell to the government establish similar relationships, in other countries as well as in the United States. The methods and focus of attention may vary according to the political system, but the basic effort is the same. Similarly, government-regulated firms in the United States are often heavily staffed with former employees of their regulatory agencies. That the military should take steps to protect itself in the bureaucratic and political arena is entirely to be expected, as is the willingness of civilian enterprises to assist. Merely regaling the reader with muck-raking tales about such acts would not prove that they inflated America's military spending much above the level necessary for national defense. In fact, without bureaucratic politics by the armed services, military needs might well be neglected. Domestic civil demands on government funds have often been insistent and not always worthy; it seems to be necessary for the most deserving of causes to play the basic political game.

A New Concern

What *is* disconcerting, however, is the size of the American military establishment, and thus the enormous impact unavoidably made whenever generals, admirals, congressmen, and businessmen combine to promote multibillion-dollar expenditures. As Kenneth Boulding likes to remind us, the Defense Department's annual budget is larger than the total GNP of all but six or seven of the world's countries. Its size is particularly disconcerting in light of the demonstrated staying power of the military establishment; that is, its ability to retain, after war, much of its wartime expansion. It is now large, not only by past American and contemporary international standards, but by comparison with almost any country in recent centuries.

Adam Smith noted that in the Europe of his time, "it is commonly computed that not more than one hundredth part of the inhabitants of any country can be employed as soldiers, without ruin to the country which pays the expense of their service."[30] One of the "benefits" of the industrial revolution, however, has been the release of governments from this economic constraint. When a small minority of the labor force suffices to provide all the agricultural products needed to sustain the populace, many men can be put into uniform or employed in arms industries without drastically lowering civilian consumption. Thus the cost of a large army, while surely substantial, can remain within "tolerable" limits even when there are no active military hostilities to encourage civilian sacrifice.

In part because we are rich then, there has until recently been little questioning of, let alone opposition to, the maintenance of a large military establishment in the United States. It has required no cut in American living standards; only giving up some increases that would otherwise probably have occurred. While Americans might have wished for fewer guns and soldiers, they have regarded the armed forces as a thoroughly necessary evil. Presidential candidates campaigned against missile gaps, and in opinion surveys a large majority of the populace consistently expressed the view that the defense budget was either about right or too low. In April 1960, a period of relative international peace, only 18 percent of Americans believed their country to be spending "too much" on defense; in 1950, just *before* the beginning of hostilities in Korea, more than six out of ten voters favored an increase in defense outlays.[31]

But now that situation has changed. For the first time during the cold war, a July 1969 American Institute of Public Opinion survey found a 52 percent majority that favored a reduction in military strength. Congress reflected this new public mood, as senators turned newly jaundiced eyes upon the defense budget. With an easing of cold war tensions and new demands on our resources, once tolerable expenses come under new scrutiny. Many Americans ask whether our armed forces are not bigger than they need to be to serve their function. Answering in the affirmative, they then demand to know what the costs are to the rest of the social and economic system. Among the alleged costs of vigilance are:

1. There is a symbiotic relationship between spending and politics; for instance, congressmen who support military expenditures also benefit disproportionately from them in their constituencies. More important, this relationship has wider effects on the political system; for example, military spending provides political support for legislators with "hard-line" positions across a spectrum of military and foreign policy issues and some domestic matters as well.

2. The United States has failed, perhaps less for a lack of effort on its own part than from the "selfishness" of its allies, to use its system of military alliances as a means of reducing the American defense burden.

3. Military spending tends to come largely at the expense not of private consumption but of investment and public expenditures for social needs such as health and education, thus endangering the long-run welfare of the country.

These are some of the matters to which we shall address ourselves in the remainder of this book. We will not find out all we would like to know about the causes and consequences of military spending, but some old questions will be answered and new ones raised. Where feasible a comparative perspective, examining the experience of other nations, will help us better to understand America's problems.

2. The Spectrum of Defense Issues

An Open Purse

The power of the purse has for centuries been the primary legislative instrument for restraining an extravagant or self-serving executive. In the United States, the Congress must both authorize and fund all Defense Department expenditures. Yet Congress's success in cutting the military budgets presented to it is unimpressive even when compared with the experience of Tudor parliaments and Henry VIII.

Former Secretary McNamara and his civilian analysts in the Office of the Secretary of Defense reduced or eliminated many armed forces programs of debated merit. Before him, both Republican and Democratic executives were often successful in imposing a ceiling on military expenditures and in forcing the services to reach an agreement on who was to get what and who would have to accept less or go without. Frequently the Bureau of the Budget is able to eliminate defense programs that seem wasteful as compared with competing civilian needs for funds. But once the executive branch has reached agreement on its defense request from Congress, not once in the past quarter-century has the legislature voted down a major weapons system proposal. It has always provided virtually everything asked of it, and sometimes more.[1] Congress rebuffed McNamara's efforts to reduce and reorganize the national guard and military reserve system; it often successfully resists the closing of army, air force, and naval bases; it has not infrequently voted funds above the request for ships, the Marine Corps, more aircraft wings, or a new manned bomber. It has not been rare for an executive to find himself trying to hold back a coalition of legislators, industrialists, and officers who seek to evade restraints being imposed by President, Budget Bureau, or Secretary of Defense. Even President Eisenhower, with his enormous prestige as judge of military needs, had this experience.

Perhaps it was not always this way. One analyst of American civil-military relations insists that in the old (pre-World War II) days of no war and no crisis, civilians in government were greatly suspicious of the armed services, and defense appropriations were very carefully scrutinized.[2] But as long ago as 1950 another scholar was already complaining about the looseness of Congress's purse when the armed services made their requests.[3] Certainly it is easy to report anecdotes of pork-barreling in Congress, where one senator supports a contract for arms purchases in another senator's state, in return for the other senator's help in preventing the Pentagon from closing a naval base in his own state. Key congressional committee members and chairmen are reputed to be especially favored by defense contracts for their districts.[4] The long-standing practice of allowing congressmen of the President's party to announce the award of defense contracts to their districts does nothing to diminish the image of military spending as contributing to legislators' political well-being.

But as we said about the military and bureaucratic politics, this is not an unusual pattern of behavior, nor censurable within the common norms of American politics. The defense budget is enormous, much larger than all other federal programs combined. Congress spends proportionately far less time analyzing the Defense Department's spending than that of any domestic agency. As an omnibus bill, the defense appropriation is a classic

candidate for the American political art of logrolling. It
can be so constructed that a majority of congressmen
will appear to have gains for their districts' incomes, at
the expense of a minority of districts that lack much
defense impact. Those who relate the horror stories of
favor-swapping misunderstand the system if they think
that merely exposing such cases will end the practice.
Especially in the American political system, where power in the legislature is highly decentralized, representatives and senators are heavily dependent on their own
local power bases among their constituents. They cannot expect the kind of support from the party central
office that in Britain helps make it possible for a law-
maker to resist an exchange of favors to further purely
local interests. It is well known that, from the point of
view of marginal costs and gains to the entire community, the pork barrel leads to excessive spending for rivers and harbors; too many VA hospitals may be built,
etc. A deductive argument has been made that because
of logrolling public expenditures in a democracy are
typically "too high."[5]

Whether or not defense appropriations are "too high,"
allowing for the failure of some technologies that one
must expect, or by comparison with other government
programs, we cannot say here. But we can investigate a
number of questions bearing on that one.

1. Is there a general pattern of voting on defense expenditures that holds up over a variety of particular
weapons systems? If so, it is clear that particular authorizations or appropriations owe little to persuasive arguments about the need for individual weapons, and much
to the general orientation of congressmen. Voting analysis in itself will not, however, enable us to distinguish a
pro-defense coalition put together by logrolling from

one based on ideology or deep-seated conviction about preparedness.

2. Is there a relationship between voting on defense expenditures and on other questions about the armed services and, more broadly, about basic political questions of arms control, disarmament, the United Nations, and East-West relations? Perhaps also, are defense appropriations related to those for NASA, or other aerospace activities such as the supersonic transport? If yes, especially to the first question, we have evidence suggesting a roughly "ideological" strain in congressional action, going beyond the mere exchange of favors in logrolling.

3. Is there indeed evidence that congressmen's voting is related to economic benefits for their districts? What is the relationship between legislative voting and patterns of DoD expenditure by state? Does this relationship hold up when we control for such other allegedly powerful influences as party, region of the country, or urbanization of the congressman's district? If yes, the logrolling hypothesis is supported. And if also yes to question number 2 above, there is a suggestion that the effect of logrolling goes well beyond maintaining high defense appropriations and extends to support for a broad cluster of political views and the legislators who hold them.

4. What has been the pattern of the above relationships over time? Is the cluster of defense-related issues becoming larger or more distinct from matters of purely domestic politics? And is any relation of voting to state-by-state expenditures becoming more, or less, close?

Scaling Senate Roll-Call Votes

We shall investigate these questions in this chapter and

in the following one. The procedure is as follows: First, we shall look at all roll-call votes in the United States Senate for the 87th and 90th congresses covering the years 1961-62 and 1967-68. Choice of these two congresses will enable us to encompass congressional politics for much of the past decade, and they are separate enough so that we may hope to find important changes over time. By 1967-68 the problem of the "military-industrial complex" had begun to attract political attention (though not as much as in 1969-70). Nevertheless, with these years we avoid complications in the analysis that would be introduced by choosing years when different parties organized the legislature or when there were partisan changes at the White House. The roll-call data are published annually in the *Congressional Quarterly Almanac,* along with reports on "pairs" by legislators and responses to *CQ* polls by many legislators who were unable to vote.

We choose the Senate rather than the House partly for convenience. When dealing with many roll-call votes, a body of 100 members is more manageable, even with the aid of a computer, than is one of 435 members. More important, however, is the availability of data on DoD and NASA expenditures by state and not by congressional district. A number of government reports and private studies, to be cited in the next chapter, deal with the states; almost none consider the defense impact at the level of the county, district, or standard metropolitan area.[6] This is unfortunate, because the political impact of defense expenditures may be even more potent in the House than in the Senate. Except for Alaska, in no state is the contribution of military spending to total personal income more than 25 percent, but there must be quite a number of congressional districts, encompassing large bases or manufacturing plants, where the 25 percent figure is exceeded. A senator can usually build his political support on a variety of groups; a representative often has little choice. But the data simply are not adequate for study of the House, and, as we shall show in the following chapter, the impact on senators of military expenditures, even on the aggregated state level, is usually strong enough to show up under examination.

Voting analysis is recommended by its use of publicly available information on important political matters. On roll-call votes senators take positions that become known to their colleagues, to other government officials, and to their constituents. The alignments are printed in major newspapers and scrutinized systematically by a number of publications (including *Congressional Quarterly* and the *New Republic*), and senators know their actions will be watched. Roll calls occur frequently enough to cover many issues and, including votes on amendments and procedure, to show relatively subtle differences of opinion. Of course many sections of bills or even entire measures never come up for roll-call votes, and, too, the aye or the nay for public consumption may conceal various shades of enthusiasm. Only intensive interviewing can hope to establish these. But for identifying positions on the broad, major issues of politics, roll-call voting analysis is unsurpassed. It has been applied by so many scholars to the study of legislative and quasi-legislative bodies, such as state and local legislatures, both houses of Congress, European parliaments, the Supreme Court, the United Nations General Assembly, and United Nations conferences, as to require no special citations or justification here.

Our basic questions concern the existence of a rather

broad spectrum of defense-related issues. Hence we are interested not in particular roll calls, where idiosyncratic influences may be serious, but in the general sweep across a number of votes. We ask whether there are one or more sets of roll calls on which the individual vote alignments are similar. In accordance with the need for searching widely, we began with all roll calls in the 87th Congress and in the 90th Congress that were substantively concerned with Defense Department authorizations or appropriations; other defense matters such as promotions of military officers, the draft, civil defense, and military aid or arms exports; problems of East-West relations such as the Consular Convention with the Soviet Union and East-West trade; arms control and disarmament matters; and questions of international organization, such as American participation in efforts to resolve the United Nations budgetary crisis in the early 1960s. To these we added appropriation or authorization votes for NASA, helicopter subsidies, airports, the supersonic transport, and the COMSAT controversy of 1963, in an effort to see whether military votes were related to civilian aerospace interests. Finally, for the 90th Congress we added the votes on gun control legislation—a nonmilitary matter but one in which many arms manufacturers would have particular interest. Both congresses were analyzed separately, but for each the two sessions were combined, making a total of over 70 roll calls for each Congress.

We then proceeded inductively, looking for recurring alignments or "Guttman scales" among the votes, where one could "predict" a senator's voting position on one roll call by his position on previous roll calls. Where a recurring alignment exists, one can say, with a low likelihood of error, that if a senator was one of only a few "doves" on a vote where the doves were isolated (e.g. an amendment to cut DoD appropriations across the board by 10 percent) he will also be dovish on roll calls where there were more doves (e.g. some anti-ABM votes). The general principles for this kind of analysis are well accepted; suffice it to say in summary that: (1) we do not specify in advance which roll calls we expect to find with similar alignments (except for the original demarcation of a large subset of all Senate votes) or how many separate sets of alignments we expect to find; (2) we produce alignments or scales on which particular roll calls, or issues, can be ranked according to the direction of some underlying dimension or superissue (hawk vs. dove, or pro- or anti-NASA, for instance); (3) also, senators can be ranked on each scale according to their position on the dimension, giving us a set of information on senators' behavior, across a number of roll calls, that can be related to their own characteristics (e.g. party membership) or attributes of their constituencies (e.g. northern vs. southern, or percentage of state income deriving from Defense Department expenditures).[7] In this chapter we shall confine ourselves to identifying the basic dimensions of the two congresses and senators' positions on those dimensions; the following chapter will discuss the distribution of DoD spending and how it is related to voting in the Senate.

Three other points about this particular application should be noted. First, when a senator did not vote in a roll call, but was "paired" for or against the measure or announced his position, either at the time or in response to a *Congressional Quarterly* poll, his action is treated as equivalent to a vote since it is part of the public record. Thus there are only three possible positions for analytical purposes—pro, anti, or absent. Absences are relatively rare (less than 10 percent of the total record); they are treated as missing data and do not enter into a decision of where on the scale a senator should be scored. Secondly, if we find two or more separate scales in a Congress the scales will not necessarily be completely independent (uncorrelated). They may be distinct but nevertheless somewhat related, and the degree of relationship will have to be checked in each case. Finally, roll-call votes with a division more extreme than 95 percent to 5 percent (among senators recorded) were dropped from the analysis. This produces a slight but entirely manageable loss of differentiation among the most extreme senators at either end of the spectrum;[8] inclusion of votes with extreme divisions would distort the computations.

Defense-related Issues in 1967-68

We shall look first at the 90th Congress, partly because it is more recent than the 87th and readers probably will remember more about the issues voted upon, and partly because there was some evolution that makes the 87th Congress's experience easier to interpret if one knows what came later. Four distinct scales emerged; their composition is shown in Table 2.1. The roll calls are listed in order of their strength on the underlying dimension. The first scale, for instance, is a generalized defense and foreign policy scale, and votes are ranked from issues on which most senators agreed, with the "hawks" isolated, to the most "dovish," with the liberals isolated. All motions were accepted except those marked "defeated." The vote number at the left corresponds to that used by *Congressional Quarterly Almanac* for the first session; second session votes are

Table 2.1. Defense-related Scales in the 90th Congress

Scale 1: *General Defense and East-West Relations* (47 items)

C.Q. No.	Substance
25	Reconvene Geneva Conference to end Vietnam war
47	Not ratify Consular Convention with Russia unless J.C.S. certify no Soviet aid to North Vietnam (defeated)
48	Add clause to Cons. Conv. about no restrictions on U.S. security (defeated)
45	No Cons. Conv. without USSR press freedom (defeated)
46	No Cons. Conv. without end to Vietnam war (defeated)
50	Ratify Cons. Conv.
400	1969 military procurement authorization—3% reduction
44	Amend Cons. Conv. eliminating criminal immunity for Soviet consular officials (defeated)
49	Honorable conclusion of war before Cons. Conv. (defeated)
475	$280,000 for Charleston, S.C., air force base (defeated)
153	Foreign assistance—expand credits for arms sales (defeated)
374	Financial penalty to countries trading with North Vietnam (defeated)
249	Suspend rules to amend Military Construction Act
394	1969 military procurement—limit R & D to $7.4 billion (defeated)
396	Bar ABM funds until system practical and cost known (defeated)
468	Bar ABM funds until July 1969 (defeated)
403	Authorize $20 million to Arms Control and Disarmament Agency (ACDA) for 1969-70 (defeated)
536	Reduce army construction by $227 million (defeated)
527	Prohibit aid to communist countries until Pueblo released
571	Bar 1969 funds for ABM production and deployment (defeated)
495	Reduce funds for civil defense (defeated)
402	Authorize $33 million to ACDA for 1969-70 (defeated)
395	Reduce authorization for missiles (defeated)
470	Reduce military construction authorization to House level (defeated)
426	Table amendment to allow individual counsel before draft boards
472	Reduce army construction funds in U.S. 10% (defeated)
106	Make conscientous objector status harder to get
471	Eliminate $17 million for army facilities in Germany (defeated)
91	Allow individual counsel before draft boards (defeated)
380	Expenditure reductions from military and space programs (defeated)
486	Ban nonmilitary sonic booms and study effects (defeated)
474	Reduce navy and air force authorization 10% (defeated)

Table 2.1 (cont.)

C.Q. No.	Substance
398	Reduce 1969 military procurement to 1968 level (defeated)
473	Reduce navy and air force authorization (defeated)
469	Eliminate 227 million for ABM (defeated)
86	Amend Selective Service Act, extend for 2, not 4 years (defeated)
576	Limit military R & D to $7.1 million (defeated)
85	Volunteer army to replace draft (defeated)
477	Reduce B-52 bomber operations in 2d supplementary appropriation for Vietnam (defeated)
573	Eliminate funds for Sage bomber defense (defeated)
90	Draft deferments for Vista and Peace Corps (defeated)
170	Recommit defense appropriation for 3.6 billion cut (defeated)
574	Reduce appropriations for military personnel 10% (defeated)
89	Require President to establish national draft criteria (defeated)
575	Bar funds for surface-to-air missiles (defeated)
168	Cut defense appropriations 10% (defeated)
573	Bar funds for procuring some military helicopters (defeated)

Coefficient of reproducibility, omitting absences   = .92
Coefficient of reproducibility, counting absences as ½ errors = .88

Scale 2: *NASA* (9 items)

459	Reduce funds for various space programs by $140 million
458	Reduce NASA authorization (defeated)
496	Increase funds for NASA facilities
218	Reduce Voyager appropriations (defeated)
217	Reduce NASA appropriations $100 million (defeated)
120	Reduce NASA authorization $100 million (defeated)
494	Reduce NASA R & D by $300,000 (defeated)
119	Reduce NASA authorization (defeated)
546	Reduce by half the minimum amount subject to renegotiation on contracts (defeated)

Coefficient of reproducibility, omitting absences   = .93
Coefficient of reproducibility, counting absences as ½ errors = .88

Scale 3: *Gun Control* (15 items)

556	Authorize import of some types of guns for sport (defeated)
577	25% ceiling on indirect costs payable on military research grants (defeated)
418	Allow mailing of handguns with affidavit (defeated)
549	Exempt some types of ammunition from control (defeated)
551	Table amendment to reconsider gun dealer qualifications
550	Tighten qualifications for licensed gun dealers (defeated)

Table 2.1 (cont.)

C.Q. No.	Substance
553	After 1970 require state gun control laws (defeated)
555	New title for license of concealable weapons (defeated)
417	Against exception on transferring destructive devices (defeated)
416	Reguire affidavit from buyer to mail gun (defeated)
552	New title for license of registered firearms (defeated)
554	New title for collecting information on all firearms (defeated)
414	Add long gun coverage to firearm control (defeated)
415	Against shipments of long guns (defeated)
149	No Export-Import Bank credit to country aiding U.S. enemies

Coefficient of reproducibility, omitting absences   = .93
Coefficient of reproducibility, counting absences as ½ errors = .90

Scale 4: *Arms Sales Abroad* (4 items)

156	Table motion to reconsider Tower amendment
155	Tower amendment promoting arms sales abroad (defeated)
154	Table Tower amendment (defeated)
147	No Export-Import Bank credit on arms sales to underdeveloped countries (defeated)

Coefficient of reproducibility, omitting absences   = .96
Coefficient of reproducibility, counting absences as ½ errors = .95

added consecutively in *CQ* order. A few senators, chiefly those who died in mid-term and their replacements, have too few recorded votes to be placed on some scales.

By far the most important scale emerging is one concerned with quite a broad spectrum of votes on defense and East-West relations. In addition to Defense Department appropriations and authorizations, there are votes on money for the Arms Control and Disarmament Agency (ACDA), on the Vietnam War, seven roll calls on the draft, aid to communist countries, and ratification of the Consular Convention with the Soviet Union. In all, 47 roll calls cluster together, including almost all of those originally identified as defense-related.

Finding virtually all of the defense appropriation and

authorization measures closely associated with each other is very important, as it suggests that in Congress particular military expenditures are rarely evaluated solely on their individual merits but rather are seen as aspects of a more general attitude toward defense spending. Senators who voted to limit military research and development in general also voted, almost without exception, against the ABM. Similarly, those who favored allocating $280,000 to a Charleston, South Carolina, air force base also virtually without exception favored the ABM. One possible explanation for this behavior is in logrolling: one senator gets funds for the military base in his state while another, in exchange for his support, gets missile-manufacturing contracts for the aerospace firm in his constituency. Another explanation is that senators have general convictions, perhaps ideological, about the merits or perils of defense spending. Each may have a different cutoff point above which he would not approve more expenditures, but on either side of that point he could rank controversial spending measures in an order of desirability that would agree generally with the rankings of other legislators.

In light of this second possibility, the fact that defense expenditure attitudes are closely linked with a wider spectrum of defense and foreign policy issues is important. Matters of the draft, a Consular Convention with the Russians, miniscule appropriations for the Arms Control and Disarmament Agency, or the Vietnam War have little to do with direct economic gains for anyone's constituents; they are not subject to logrolling in the same way as are military expenditures. This indicates, therefore, that the logrolling explanation, though perhaps partially correct, is incomplete. Senators appear to view quite a wide range of issues in much the same manner, and their attitudes on all are interrelated. Whatever the importance of logrolling—and that can be better evaluated in the next chapter when we relate defense voting to Defense Department expenditures by state—more general convictions are also operating.

Moreover, the presence of a linked spectrum of defense and foreign policy issues is unusual by comparison with the general population of Americans. Even among college-educated citizens, Davis Bobrow found in national surveys in 1963 and 1964 that *"Continental defense opinions are not strongly associated with hypothetical dispositions to value different international futures or to perceive different patterns of agreement on disarmament. . . .* Accordingly, we were not justified in assuming that preferences for particular international futures have any clear implications for the perceived desirability of weapons systems." He added that attitudes on ABM fragment into a variety of small clusters of subissues, "and therefore we cannot predict from one opinion about active defense to many other opinions about active defense."[9] I hesitate to stress these differences between Congress and populace too heavily; after all, one would expect more structured opinions among elites constantly involved in political decision-making, and, as we shall see below, senators' opinions were less structured in the early 1960s than they became later. Nevertheless, there remains a suggestion that some special influence—possibly political pressures generated by defense spending—has helped senators develop a more tightly knit perspective on this range of issues.

Several other distinct scales also appear, though none is as pervasive as the general defense and East-West relations dimension. The space program roll calls, largely for NASA, cluster surprisingly well together, separate from

defense spending issues. Recall that this result emerged
inductively from the analysis; we originally made no
specifications as to whether any particular votes should
be grouped together. The separateness of NASA questions from military ones is relevant to statements about
a military-industrial defense and aerospace complex that
allegedly pushes up expenditures in these areas. The defense and civilian aerospace programs are not identical
in this sense, and logs are not rolled between them with
the same facility that they perhaps may be rolled within
either program. Despite the diversity of the various military purchases—for bases, conventional weapons, ships,
missiles, and aircraft—there is a generalized perception
as to their desirability, a perception that differs somewhat from that most senators hold about the civilian
space effort. It is quite common for senators to favor
one and oppose the other. Almost certainly, some ideological views on communism, preparedness, or fear of
war that influence votes on defense issues do not extend
equally to the space program for many legislators.

Similarly, gun control seems to be a separate dimension, where the Senate alignments again are different
from those on defense and space. The votes on gun
control hang together in a scale that includes only two
roll calls from all those connected with defense or East-
West relations. Again, if there is a "military-industrial
complex" promoting military expenditures, just as it
does not fully encompass the civilian space program, it
does not extend perfectly to the manufacture and sales
of small arms. Other attitudes and pressures—on hunting, on the individualist tradition of self-defense, perhaps on race relations—are at issue.

Finally, the separateness of four roll calls on arms
sales abroad is notable. The mix of arms to be sold

abroad is of course different from that purchased for the American armed forces, but the latter is a great enough conglomerate of conventional and aerospace weapons to include, among others, virtually everything for foreign sale. It is thus interesting that the proponents of arms for underdeveloped countries are not quite the same as those who most favor arms for the United States. Again, ideology on defense and foreign policy, including foreign aid, is present along with any general desire to help arms-makers.

Thus we have several distinct scales. The size and composition of the defense and East-West relations scale is

Table 2.2. Rank-order Correlations (Tau) among Defense-related Scales in the 90th Congress

	General Defense	NASA	Gun Control	Arms Sales
General Defense	1.00	.30*	.41*	.31*
NASA		1.00	.16	.28*
Gun Control			1.00	.15
Arms Sales				1.00

*Statistically significant at the .001 level.

probably the most important finding, but the failure of some other seemingly related issues to be closely associated with it must not be ignored. This failure can, however, be exaggerated, since the scaling method employed did not demand completely separate and uncorrelated scales. In fact there are moderately strong relations among many of the scales, as can be seen in table 2.2.[10]

While still distinct, some relation between senators' stances on, for example, gun control and general defense issues is nevertheless apparent. The two sets of attitudes, though not close enough to form a single scale, are not independent either. The same is true in somewhat less degree for a relation between general defense and the other two scales, for NASA and arms sales abroad. The United States military space program, in DoD, is roughly one-fourth as large as NASA;[11] hence one would expect that, on economic as well as perhaps ideological grounds, some supporters of defense spending would also be NASA advocates. A systematic examination of the correlates of each set of attitudes must wait for the next chapter, but some better sense of both the distinctness and the relations among these sets of issues can be had by looking at the rankings of individual senators. Table 2.3 lists them in order, with the scale position noted to the left of each list, in descending order from what would popularly be considered the "conservative" to the "liberal" position. The rank numbers indicate the number of roll calls separating each position from the one immediately preceding it. For example, on the general defense scale the first four ranks are distinguished from one another by only a single roll call; the positions taken by Senators Miller and Holland, however, differed on four roll calls as indicated by their adjacent scores of 3 and 7 respectively. While it is useful to have this information, the scores should not be taken as absolute values on an interval scale. They are too heavily influenced by the frequency with which particular issues may have happened to meet with roll-call votes in a particular session. In the analysis we will look only at the rankings, considering Senators Miller and Holland as neither more nor less distinct on this scale than Tower and Dodd.

There are some similarities among the scales. For instance, Senators Curtis, Murphy, and Thurmond all are recorded at the most "conservative" position on all four. This is the meaning of the moderately high correla-

Table 2.3. Senators' Scores on 90th Congress Defense-related Scores

1 General Defense		2 NASA	3 Gun Control	4 Arms Sales
0 Cotton	(N.H.)	0 Allott	0 Curtis	0 Allott
Curtis	(Neb.)	Anderson	Hansen	Anderson
Dominick	(Colo.)	Bartlett	Hickenlooper	Baker
Eastland	(Miss.)	Bible	Hruska	Bennett
Fannin	(Ariz.)	Cannon	Murphy	Boggs
Hansen	(Wyo.)	Curtis	Thurmond	Brooke
Hill	(Ala.)	Dodd	1 Allott	Cannon
Hollings	(S.C.)	Eastland	Baker	Curtis
Hruska	(Neb.)	Ellender	Carlson	Dirksen
McClellan	(Ark.)	Fannin	Dirksen	Dodd
Murphy	(Cal.)	Hayden	Dominick	Dominick
Russell	(Ga.)	Hill	Fannin	Eastland
Stennis	(Miss.)	Holland	Hatfield	Fannin
Talmadge	(Ga.)	Jackson	L. Jordan	Fong
Thurmond	(S.C.)	Jordan	Morse	Griffin
Tower	(Tex.)	Kuchel	Prouty	Hansen
1 H. Byrd, Jr.	(Va.)	E. Long	Sparkman	Harris
Dodd	(Conn.)	R. Long	Tower	Hart
Ervin	(N.C.)	Magnuson	M. Young	Hayden
Mundt	(S.D.)	McClellan	2 Bennett	Hickenlooper
M. Young	(N.D.)	McGee	Bible	Hill
2 L. Jordan	(Ida.)	Metcalf	Burdick	Hollings
3 R. Byrd	(W.Va.)	Monroney	R. Byrd	Hruska
Lausche	(Ohio)	Montoya	Cannon	Inouye
Miller	(Ia.)	Murphy	Cotton	Jackson
7 Holland	(Fla.)	Pearson	Eastland	Javits
Montoya	(N.Mex.)	Percy	Ervin	B. Jordan
8 Allott	(Colo.)	Smathers	Hill	E. Kennedy
Anderson	(N.Mex.)	Smith	Hollings	Kuchel
Baker	(Tenn.)	Sparkman	B. Jordan	Lausche
Bennett	(Utah)	Stennis	R. Long	R. Long
Carlson	(Kans.)	Thurmond	McGovern	Magnuson
B. Jordan	(N.C.)	Tower	Metcalf	McGee
Smathers	(Fla.)	Yarborough	Miller	Miller
Smith	(Me.)	Young	Moss	Mondale
9 Dirksen	(Ill.)	1 Aiken	Mundt	Montoya
Fong	(Ha.)	Carlson	Russell	Morton
Hayden	(Ariz.)	Dominick	Stennis	Murphy

Table 2.3 (cont.)

	1 General Defense		2 NASA	3 Gun Control	4 Arms Sales
9	Hickenlooper	(Ia.)	1 Fong	2 Talmadge	0 Muskie
	Jackson	(Wash.)	Hickenlooper	3 Bartlett	Pastore
	Magnuson	(Wash.)	Hruska	H. Byrd, Jr.	Pearson
	Sparkman	(Ala.)	Inouye	Church	Percy
10	Cannon	(Nev.)	E. Kennedy	Ellender	Prouty
11	Bible	(Nev.)	Mondale	Fulbright	Scott
	R. Long	(La.)	Mundt	Gruening	Sparkman
	Spong	(Va.)	Muskie	McClellan	Stennis
13	Boggs	(Del.)	Scott	McGee	Talmadge
	Brewster	(Md.)	2 Hartke	Muskie	Thurmond
	Griffin	(Mich.)	Jordan	4 Aiken	Tower
	Harris	(Okla.)	Mansfield	Holland	3 Case
	Kuchel	(Cal.)	McCarthy	6 Boggs	Hartke
	McIntyre	(N.H.)	McIntyre	Harris	R. Kennedy
	Pearson	(Kans.)	Prouty	Montoya	Mansfield
14	Bayh	(Ind.)	3 Symington	Pearson	McCarthy
	Ellender	(La.)	5 Baker	Yarborough	McIntyre
	McGee	(Wyo.)	Bennett	7 Anderson	Metcalf
	Monroney	(Okla.)	Brooke	Jackson	Monroney
	Pastore	(R.I.)	Dirksen	Lausche	Moss
16	J. Williams	(Del.)	Ervin	Magnuson	Randolph
18	Ribicoff	(Conn.)	Javits	Spong	Ribicoff
19	Aiken	(Vt.)	Morton	8 Mansfield	M. Young
20	E. Long	(Mo.)	Ribicoff	10 Bayh	4 Aiken
	Morton	(Ky.)	7 R. Byrd	12 Hart	Bartlett
	Muskie	(Me.)	M. Young	Scott	Bayh
	Pell	(R.I.)	8 Clark	13 Hartke	Bible
	Prouty	(Vt.)	Griffin	Randolph	Brewster
	Randolph	(W.Va.)	Hansen	J. Williams	Burdick
	Scott	(Penn.)	Harris	14 Brewster	H. Byrd, Jr.
23	Bartlett	(Alas.)	Hatfield	Dodd	R. Byrd
	Inouye	(Ha.)	Moss	Fong	Carlson
	Mansfield	(Mont.)	Randolph	Griffin	Church
	Metcalf	(Mont.)	Spong	McIntyre	Clark
	Symington	(Mo.)	Talmadge	Nelson	Cooper
24	Cooper	(Ky.)	H. Williams	Proxmire	Cotton
	Gore	(Tenn.)	9 Bayh	Ribicoff	Ellender
	Percy	(Ill.)	Boggs	Smith	Ervin

Table 2.3 (cont.)

1 General Defense		2 NASA	3 Gun Control	4 Arms Sales
24 H. Williams	(N.J.)	9 Brewster	14 Symington	4 Fulbright
28 Javits	(N.Y.)	Burdick	15 Brooke	Gore
29 Brooke	(Mass.)	H. Byrd, Jr.	Case	Gruening
McCarthy	(Minn.)	Case	Clark	Hatfield
31 Case	(N.J.)	Church	Cooper	Holland
32 Burdick	(N.D.)	Cooper	Goodell	L. Jordan
Mondale	(Minn.)	Cotton	Gore	E. Long
Moss	(Utah)	Fulbright	Hayden	McClellan
Tydings	(Md.)	Gore	Inouye	McGovern
35 E. Kennedy	(Mass.)	Gruening	Javits	Morse
R. Kennedy	(N.Y.)	Hart	E. Kennedy	Mundt
37 Church	(Ida.)	Hollings	Kuchel	Nelson
Yarborough	(Tex.)	R. Kennedy	Mondale	Pell
38 Hartke	(Ind.)	Lausche	Pastore	Proxmire
40 McGovern	(S.D.)	McGovern	Pell	Smathers
Nelson	(Wisc.)	Miller	Percy	Smith
Proxmire	(Wisc.)	Morse	Smathers	Spong
43 Hart	(Mich.)	Nelson	Tydings	Symington
44 Hatfield	(Ore.)	Pastore	H. Williams	Tydings
46 Gruening	(Alas.)	Pell	S. Young	H. Williams
S. Young	(Ohio)	Proxmire		J. Williams
47 Clark	(Penn.)	Russell		Yarborough
Fulbright	(Ark.)	Tydings		S. Young
Morse	(Ore.)	J. Williams		

tions in table 2.2. But some of the differences in the rankings are quite striking. Senators Edward Kennedy and Yarborough, for example, are far toward the "dovish" or "liberal" end of the list on the general defense scale but are near or in the most pro-NASA group. (As we shall see, both are from states that gain substantially from NASA expenditures.) On the other hand, several southern senators who supported the "hawkish" or "conservative" side on virtually all the defense votes were nevertheless strongly in favor of cutting back

NASA's effort. It is, in fact, this selective impact of preparedness and fiscal economy that makes the popular labels of "liberal" and "conservative" of such dubious utility here. That difficulty is inherent in the finding that the scales were not highly correlated and hence do not well reflect a single overall liberal-conservative dimension.

Similarly, gun control is distinct, perhaps most dramatically in the case of Connecticut's Senator Dodd who, in spite of his position at the top of the other scales, has been a leader in the gun control movement. This also despite the fact that Connecticut has a very high income from Defense Department spending and, in Marlin and Winchester Arms companies, major gun manufacturers.[12] And the strongest votes against encouraging American arms sales abroad were cast by a coalition of foreign policy liberals and fiscal conservatives.

Defense-related Issues in 1961-62

The 87th Congress, six years earlier, shows important similarities and differences when compared with recent experience. Like the 90th Congress, it had a Democratic majority in both houses and a Democrat in the presidency. But the occupants of the White House were different and so in many ways were the issues and the mood of the country. In the earlier years many feared a missile gap and wanted a prestigious accomplishment in space; by the late 1960s anything remotely military risked wide unpopularity.

A general defense and East-West relations scale existed in 1961-62 too, but there were far fewer specifically defense-oriented roll calls than in 1967-68. Although there again happen to be 47 items in the scale, a ma-

Table 2.4. Defense-related Scales in the 87th Congress

Scale 1: *General Defense, East-West Relations, and COMSAT* (47 items)

CQ No.	Substance
179	Establish ACDA
227	No U.S. economic aid to countries in arrears on U.N. assessments (defeated)
147	Deny Development Loan funds to nations 2 years in arrears on U.N. assessments (defeated)
228	Amend authorization for payments on U.N. Congo operation (defeated)
423	Pass $4.4 billion military and economic aid bill, and $2.4 billion other
35	Bar aid to communist countries in arrears on U.N. assessments (defeated)
176	Delete provisions requiring SBA, DoD, and GSA to favor small business in government subcontracts (defeated)
267	Surplus food for Poland, Yugoslavia
92	Eliminate presidential authority to sponsor exchange programs with U.N. (defeated)
178	Send ACDA bill to committee for study (defeated)
36	Pass Mutual Defense Assistance Control Act
177	Eliminate ACDA authority for disarmament research (defeated)
135	Limit U.S. contribution to U.N. Congo operation to 40% of total (defeated)
268	No aid to countries that export war articles to Russia (defeated)
146	Bar assistance to countries that export arms to Soviet-controlled countries (defeated)
416	Bar aid to communist countries
415	Bar aid to countries trading with Cuba
318	No surplus agriculture aid to communist countries
315-17	28 items in the COMSAT debate
319-27	(315 and 316 defeated, almost all other votes were
329-33	to table amendments offered by COMSAT opponents)
335-46	
293	Study impact of space efforts (defeated)

Coefficient of reproducibility, omitting absences = .95

Coefficient of reproducibility, counting absences as ½ errors = .93

Scale 2: *Military Assistance Reductions* (4 items)

138	Reduce military assistance authorization $250 million
418	Reduce military assistance $150 million (2d sess.) (defeated)

Table 2.4 (cont.)

C.Q. No.	Substance
133	Reduce military assistance $500 million (defeated)
189	Reduce military assistance appropriation $150 million (defeated)

Coefficient of reproducibility, omitting absences = .95
Coefficient of reproducibility, counting absences as ½ errors = .93

Scale 3: *Aerospace and Civil Defense* (10 items)

417	President must publish in Federal Register reasons for aiding a communist country
109	Delete $12 million research funds for supersonic transport (defeated)
113	Cut $2.5 million in OCDM radiological defense equipment (defeated)
110	Table amendment to cut $50 million from Independent Offices bill (incl. aerospace and civil defense)
111	Delete $6.9 million in CAB helicopter subsidy (defeated)
363	Send NASA appropriation to committee to cut $275 million (defeated)
168	Restrict Federal Airport Act spending authority (defeated)
169	Reduce airport appropriations by $114 million (defeated)
364	Pass civil defense appropriation
362	Reduce NASA appropriation $105 million (defeated)

Coefficient of reproducibility, omitting absences = .96
Coefficient of reproducibility, counting absences as ½ errors = .93

jority actually come from the debate and filibuster on the proposals for a private corporation to operate communications satellites in space (COMSAT). A small band of about a score of senators, all but one of them Democrats, opposed creation of a private corporation with privileges that might lead to very substantial profits and fought to have the job done by a public agency. While economic benefits for a portion of the aerospace industry were indeed at stake, communications rather than aerospace firms were more directly concerned, and the defense component was minimal. Otherwise the composition of the scale is not very different from the content of the general scale in the 90th Congress, including issues such as ACDA and foreign aid to communist countries. The measures directed against countries in arrears in their U.N. assessments included some under-developed nations in their targets, but were chiefly aimed at communist states. Proponents of these mea-sures were not supporting a greater supranationalist role for the U.N. There were, however, several measures (items 228, 92, and 135) where support for the world organization was at issue—something that did not arise in the 90th Congress.

But strikingly absent from this scale are votes on De-fense Department appropriations. Although there were a very few such votes that did not scale, including ones with funds for a manned bomber, DoD appropriations were largely noncontroversial and opposed in roll-call votes by fewer than five senators, if by any. Military expenditures had not yet become matters of major de-bate, so the difference in composition of this scale from that in the 90th Congress is a major finding. Only since the early 1960s has a major dimension of voting ex-plicitly concerned with defense matters emerged in the Senate. Before then it was at best a latent concern, overtly manifested only in other issues of foreign policy and some quasi-domestic matters like COMSAT.[13]

There are two other scales, each apparently related in content to scales in the 90th Congress. One, concerning attempts led largely by Senator Ellender to reduce funds for foreign military assistance, may have some relation to the votes on arms sales abroad; another, labeled "Aerospace and Civil Defense," seems to be the prede-cessor to the later set of votes on NASA. It is a rather polyglot scale, however, including one vote on aid to communist countries, some on civil defense (a military matter), only two on NASA (then still rather a fledgling agency), one for the supersonic transport, and several for airports and other kinds of aircraft subsidies. In this Congress its substance was more aero than space. Gun control was not an issue in the days when all the political Kennedys were still alive, so there is no scale for it.

The relation of these scales to each other and to those in the 90th Congress will be clearer from table 2.5. Cor-

Table 2.5a. Rank-order Correlations (Tau) among Defense-related Scales in the 89th Congress

	General Defense	Military Assistance	Aerospace
General Defense	1.00	−.31*	−.35*
Military Assistance		1.00	.24*
Aerospace			1.00

*Statistically significant at the .001 level.

Table 2.5b. Rank-order Correlations (Tau) of Scales over Time

	87th Congress Scales		
90th Congress Scales	General Defense	Military Assistance	Aerospace
General Defense	*.63**	−.22	.17
Arms Sales	.26*	*11*	.02
NASA	.13	.04	*.20*
Gun Control	.30*	−.28*	−.19

*Statistically significant at the .001 level.

relations of senators' scores over time are of course only for those approximately 80 legislators who held seats in both congresses. Table 2.6 shows the scores of individual senators.

Unlike the situation in the 90th Congress, in the 87th the general defense scale was completely unrelated, in a positive way, to the alignments on (largely civilian) aerospace spending and on foreign military assistance. In

Table 2.6. Senators' Scores on 87th Congress Defense-related Scales

1 General Defense		2 Military Assistance	3 Aerospace
0	Bottum (S.D.)	0 Allott	0 Bartlett
	H. Byrd (Va.)	Boggs	Dodd
	Curtis (Neb.)	R. Byrd	Eastland
	Eastland (Miss.)	Carroll	Ellender
	Goldwater (Ariz.)	C. Case	Hart
	Hruska (Neb.)	Chavez	Holland
	L. Jordan (Ida.)	Dirksen	McCarthy
	Murphy (N.H.)	Dodd	McGee
	Russell (Ga.)	Douglas	McNamara
	Stennis (Miss.)	Fulbright	Neuberger
	Thurmond (S.C.)	Goldwater	Pell
	Tower (Tex.)	Gore	S. Smith
1	Butler (Md.)	Hart	Stennis
	Capehart (Ind.)	Hartke	1 Aiken
	Cotton (N.H.)	Hayden	Allott
	Miller (Ia.)	Hickey	Bible
	Robertson (Va.)	Humphrey	Burdick
2	Bennett (Utah)	Jackson	R. Byrd
	Dirksen (Ill.)	Kefauver	C. Case
	J. Williams (Del.)	Kuchel	Carroll
	M. Young (N.D.)	E. Long	Chavez
3	Beall (Md.)	O. Long	Clark
	Johnston (S.C.)	Mansfield	Cooper
	Mundt (S.D.)	Metcalf	Dirksen
4	Lausche (Ohio)	McCarthy	Engle
	McClellan (Ark.)	McGee	Ervin
	Talmadge (Ga.)	McNamara	Fong
5	Allott (Colo.)	Moss	Gruening
	Bush (Conn.)	Muskie	Hartke
	Scott (Penn.)	Neuberger	Hayden
6	Kuchel (Cal.)	Pastore	Hill
	M. Smith (Me.)	Randolph	Humphrey
7	Ellender (La.)	Scott	Jackson
	Ervin (N.C.)	M. Smith	Javits
	Fong (Ha.)	S. Smith	Johnston
	B. Jordan (N.C.)	Sparkman	B. Jordan
9	Cannon (Nev.)	Symington	Kefauver
	Carlson (Kans.)	Tower	Kerr
	Holland (Fla.)	Wiley	Kuchel

Table 2.6 (cont.)

1		2	3
General Defense		Military Assistance	Aerospace
9 Pearson	(Kans.)	0 H. Williams	1 O. Long
Prouty	(Vt.)	1 Aiken	R. Long
10 Bible	(Nev.)	Beall	Magnuson
Boggs	(Del.)	Bridges (N.H.)	Mansfield
Smathers	(Fla.)	Bush	Metcalf
11 Dodd	(Conn.)	Clark	Monroney
Morton	(Ky.)	Engle	Moss
12 Hickenlooper	(Ia.)	Hill	Muskie
13 Proxmire	(Wisc.)	Javits	Pastore
15 E. Long	(Mo.)	Keating	Prouty
17 Anderson	(N.Mex.)	Monroney	Randolph
R. Byrd	(W.Va.)	Prouty	Schoeppe
Engle	(Cal.)	Robertson	Scott
Hartke	(Ind.)	Saltonstall	Sparkman
Hayden	(Ariz.)	Stennis	H. Williams
Hickey	(Wyo.)	Thurmond	Yarborough
Hill	(Ala.)	2 Carlson	M. Young
Keating	(N.Y.)	Cooper	S. Young
Magnuson	(Wash.)	Lausche	2 Fulbright
Randolph	(W.Va.)	Yarborough	Gore
Symington	(Mo.)	S. Young	Russell
Wiley	(Wisc.)	3 Fong	Saltonstall
18 Aiken	(Vt.)	Gruening	Smathers
C. Case	(N.J.)	Holland	M. Smith
Chavez	(N.Mex.)	Kerr	3 Cannon
Cooper	(Ky.)	Smathers	4 Beall
Fulbright	(Ark.)	J. Williams	Bush
Humphrey	(Minn.)	4 Anderson	Butler
Kerr	(Okla.)	Bartlett	Capehart
Mansfield	(Mont.)	Bennett	Carlson
McCarthy	(Minn.)	Bible	Hickey
McGee	(Wyo.)	Burdick	Keating
Metcalf	(Mont.)	Butler	McClellan
Monroney	(Okla.)	H. Byrd	Robertson
Pastore	(R.I.)	Cannon	5 Anderson
Saltonstall	(Mass.)	Capehart	Boggs
H. Williams	(N.J.)	F. Case (S.D.)	Church
19 Muskie	(Me.)	Church	Cotton
S. Smith	(Mass.)	Cotton	Dworshak

Table 2.6 (cont.)

1 General Defense		2 Military Assistance	3 Aerospace
19 Sparkman	(Ala.)	4 Curtis	5 Goldwater
23 Jackson	(Wash.)	Dworshak (Ida.)	Hruska
Pell	(R.I.)	Eastland	E. Long
26 S. Young	(Ohio)	Ellender	Morse
31 Javits	(N.Y.)	Ervin	Morton
37 Church	(Ida.)	Hichenlooper	Symington
38 Hart	(Mich.)	Hruska	Talmadge
44 O. Long	(Ha.)	Johnston	Thurmond
46 Burdick	(N.D.)	B. Jordan	7 Hickenlooper
Gruening	(Alas.)	R. Long	Miller
R. Long	(La.)	Magnuson	8 Bennett
McNamara	(Mich.)	McClellan	Mundt
Moss	(Utah)	Miller	9 Tower
Yarborough	(Tex.)	Morse	10 H. Byrd
47 Bartlett	(Alas.)	Morton	F. Case
Carroll	(Colo.)	Mundt	Curtis
Clark	(Penn.)	Pell	Douglas
Douglas	(Ill.)	Proxmire	Lausche
Gore	(Tenn.)	Russell	Proxmire
Kefauver	(Tenn.)	Schoeppe	Wiley
Morse	(Ore.)	Talmadge	J. Williams
Neuberger	(Ore.)	M. Young	

fact, they were moderately negatively related; that is, the proponents of a "hard-line" position on defense issues tended to oppose the other two kinds of spending. These other two scales are slightly positively related to each other, and well-known "fiscal conservatives" such as Harry Byrd (Sr.) and other southerners are very noticeable at the bottom of both lists.

Over time, the very substantial continuity in the major general defense scale is the most impressive finding. The correlation of .63 is very high, and one tends to find the same senators at the extremes of both lists. For example, of the four men with 47 scores in the 87th Congress who sat again in the 90th, two (Fulbright and Morse) again scored 47 in 1967-68. And of eight repeaters from the 87th Congress zero category, seven again scored zero in the 90th. This impressive continuity shows that the opposition to increased military spending which surfaced in the late 1960s was latent among senators in the first years of the decade. The senators who sought to maintain economic relations with communist countries and to support ACDA and the U.N. in 1961-62 are the same ones who continued to seek improved East-West relations in 1967-68 and, in addition, began to move against high DoD expenditures in a way they had not done before. And their opponents were largely the same in both cases.[14] This interpretation of defense as an emergent major issue is strengthened by the intercorrelations among the defense-related scales in 1967-68, an intercorrelation that was totally absent in the earlier Congress.

Little continuity is apparent with the other two pairs of scales. The "military assistance" and "arms sales" scales correlate only .11 with each other, and of course the substance is not exactly the same. Nor is the substance quite the same with the "NASA" and "aerospace" scales, which correlate only .20. NASA is not prominent among the 87th Congress roll calls, and the aviation industry and civil defense votes are. The emergence of NASA as a major money-consuming federal agency may be very important here, as may the regional pattern of expenditures that emerged. By the 90th Congress both Texas senators, Tower and Yarborough, were numbered among NASA's strongest champions. But in the 87th Congress Yarborough was only in the second-highest rank, and Tower was virtually at the bottom of the scale. More on this in the next chapter when we correlate senators' scores with their districts' receipts from federal expenditures.

3. Defense Spending and Senatorial Behavior

Some Expectations about Defense Politics

One expects senators to support, or at least not to oppose, federal spending that directly benefits their constituents. That phenomenon alone would hardly deserve detailed study here. And according to the logrolling principle of politics, senators often support certain kinds of spending even though it does not directly benefit their constituencies. They do so in exchange for other senators' support for expenditures that do help their states. The scope of the logrolling may be wide or relatively narrow, hence the particular exercise of logrolling may be limited to defense and space appropriations, or to a subset thereof, or extended to include other kinds of spending such as highways or rivers and harbors. It may be extended further to include the exchange of other kinds of favors on legislation or appointments.

Because of the high correlations among almost all roll calls dealing with defense expenditures (whether from logrolling or other influences is not clear), we can readily go on to investigate the relation between defense spending by state and senators' votes on virtually the entire group of defense expenditure roll calls. Furthermore, since other defense and foreign policy issues are closely related to the spending roll calls, we can investigate the relation between state-by-state spending patterns and that broader spectrum. We cannot, of course, easily establish a cause-and-effect relationship nor readily draw a boundary around the defense-related set of roll calls and say that the effect ends there. At best we can establish that certain other specific roll calls or classes of votes are not related to spending patterns.

Regarding specifically defense-related issues—making up the general defense scales of the previous chapter—the following expectations seem reasonable:

1. Senators representing states where defense spending is particularly heavy will be high on the "pro-spending," "pro-defense," "hawkish" end of the scale in both congresses. They may be so because of the financial support available from the industries involved or the potential electoral support from both the public and private defense-employed sectors in their states. Or, particularly among members of the same party as the executive, they may be rewarded for past, present, or anticipated support for military spending by having contracts or military installations placed in their states. Most likely the causal connection usually works both ways. Heavy defense-generated employment, especially at military bases, may create electoral support for a "hawkish" stance on a wider spectrum of issues than merely defense spending questions.

2. Some other senators, for reasons of patriotism, ideology, party loyalty, regional considerations, or log-rolling of wider scope, will also be at the hawkish end of the scale even though their states do not benefit from an unusually high defense-generated income. Also, the political activities of defense-oriented industries may not be limited to legislators representing the states where the industries are located. Those industries, especially large contractors, may make campaign contributions or offer other assistance to sympathetic senators or to members of key congressional committees, regardless of state of origin.

3. Senators from states with relatively little defense-generated income will be found toward the "anti-

spending," "dovish" end of the scale. Specifically on spending matters, they will tend to vote for defense reductions in general, since their states pay out in taxes for military spending more than they receive back, and there is little defense-derived actual or potential electoral support in their states. There will probably be exceptions to this pattern because of the influences suggested under item number 2, and perhaps because in some instances senators may expect to have their states rewarded in the future, by bases or by contracts, in return for present "hawkishness." But a dovish option will be open to senators from defense-poor states because they will have little directly to lose from such a stance.

4. "Doves" from defense-rich states will be rare. When they are found, we will expect them to be voted out of office in later elections or the level of defense spending in the state to decline in later years.

5. Additionally, many of the same considerations point to similar predictions regarding defense spending and support for foreign military assistance or opposition to gun control.

6. Similarly, the distribution of NASA spending by state will be related to congressional support for space ventures.

Just because these are "reasonable" expectations, however, does not mean that they will in fact hold true. We will devote this chapter to investigating them, looking both at the relationships within the span of a single Congress and changes over the period of the 1960s—and we will find some surprises. But first we need good information on spending patterns. Also, we must keep various components of defense spending separate to the degree permitted by the form in which they are reported. There is no reason to expect that legislators' behavior is influenced in the same way by having large numbers of soldiers and civilians stationed at military bases in their states as by having their states as the loci of defense manufacturing in, say, the aerospace industry. The effect may be much the same, but it should not be assumed at this stage.[1]

The Distribution of Defense and Space Expenditures

A variety of data on defense and space expenditures by state is available. The most readily accessible are those published annually in the *Statistical Abstract* on Department of Defense contract awards, payroll of civilians employed directly by DoD, and the DoD military payroll.[2] These figures extend over the entire decade of the 1960s (through 1968 as of this writing) and are good for maintaining continuity in the analysis. The major difficulty is with their limitation to direct employment and prime contracts, ignoring subcontracts. Subcontracting is often very substantial, when the prime contractor is not equipped to produce all the varied components of the final product, and is not necessarily confined to the region of the prime contractor. In addition, "A prime who subcontracts 50 percent does not produce the remaining 50 percent himself . . . Other inputs along with those officially 'subcontracted' will be purchased, and these are part of the indirect impact."[3] Other less serious difficulties include decisions as to where to count a contract with a firm that has plants in several states (typically it is counted toward the state with the firm's head office), and interstate transfers to home bank accounts (most important for military payrolls, where soldiers' homes are often far from their bases of duty).

The latter qualifications are probably not important, but the lack of information on subcontracting may well be serious.

The full data set on subcontracting patterns is kept classified by the Defense Department, because its release might disclose proprietary business information as well as for conventional reasons of military secrecy. But several unclassified studies have attempted by various means to impute the changes in the state-by-state pattern produced by subcontracting. They include estimates by economist Roger Bolton that we shall use for 1961 and 1962; by an Independent Study Board for the Department of Commerce for 1963 and 1966; and a study for 1966 by Wilbur A. Steger.[4] None of these estimates is flawless, as the authors are the first to recognize, and they are not in most cases for precisely the years—1961-62 and 1967-68—of our voting data, but they do give some indication of how much and where knowledge of subcontracting patterns would change states' ranks in any ordering of DoD beneficiaries.

Abstractly, it is not clear what the probable political effect of subcontracting may be. The initial contractor makes a profit even on that portion of his award that he subcontracts, so that must be taken into account, especially when imputing political activities to management or stockholders. In addition, the prime contractor is likely to be more conscious of the DoD origin of his work than would a secondary or tertiary contractor who fills orders from many kinds of firms which ultimately sell to a variety of customers. On the other hand, when the DoD origin of the initial order is recognized, both management and labor and indeed sellers to them in the local labor markets can be expected to take a direct political interest in defense spending.

There is also a set of data put out by the Department of Defense concerning not payroll and contracts but "defense-generated employment." It includes all military personnel and civilian employment at bases, plus employment on all DoD contracts exceeding $10,000. Only a small portion of subcontracts are covered, however, to an amount estimated at roughly 12 percent of total indirect employment.[5]

Finally, there are good data on the impact of NASA spending. Information on NASA prime contracts is available for all years from NASA publications, as are data on subcontracts for 1967 and 1968. In the latter case, unlike defense spending, the state prime contract totals are adjusted to exclude the value of contracts let outside their state by the prime contractors but to add that of all subcontracts coming in. These figures are derived from NASA's annual report; for earlier years official data on subcontracting were not published, though they have been put together for January 1, 1962, through June 30, 1963, in a special study by Gerald J. Karaska.[6]

Each of these sets must be adjusted for the size of states before they can be used in this analysis. We would expect the political impact, to the degree that there is one, to vary not with the total defense input to a state, but with the proportion of income or employment in the state that is derived from defense spending. So the state-by-state totals have been divided by total personal income (or, in the case of employment figures, by the total labor force) to get a measure of proportionate impact. The relation of these new measures to each other is shown in table 3.1. Note that pensions and other veterans benefits are not included—they represent federal obligations derived

Table 3.1. Rank-order Correlations (Tau) among
Various Measures of Defense Impact

1961 and 1962 Combined

1. S.A., Contract Awards, and Bolton, Defense Procurement   =   .65
2. S.A., Military Payroll, and Bolton, Military Wages   =   .84
3. S.A., Civilian Payroll, and Bolton, Civilian Wages   =   .88
4. Bolton, Total Defense, and ISB, Defense Impact   =   .65
5. NASA Prime Contracts and NASA Adjusted   =   .42
6. S.A., Contract Awards, and S.A., Military Payroll   =   .08
7. S.A., Contract Awards, and S.A., Civilian Payroll   =   .20
8. S.A., Military Payroll, and S.A., Civilian Payroll   =   .45
9. NASA Prime Contracts and S.A., Contract Awards   =   .30
10. NASA Adjusted and Bolton, Defense Procurement   =   .65
11. All 1961 figures correlated with same index for 1962   >   .80

except

12. S.A., Contract Awards, 1961 with 1962   =   .61
13. NASA Prime Contracts, 1961 with 1962   =   .78

1967 and 1968 Combined

14. S.A., Contract Awards, and DoD, Civilian Employment   =   .50
15. S.A., Military Payroll, and DoD, Military Employment   =   .88
16. DoD, Civilian Employment, and ISB, Defense Impact   =   .65
17. DoD, Civilian Employment, and Steger, Defense Priority One   =   .48
18. ISB, Defense Impact, and Steger, Defense Priority One   =   .62
19. Steger, Defense Priority One, and S.A., Contract Awards   =   .66
20. S.A., Civilian Payroll, and DoD, Civilian Employment   =   .61
21. NASA Prime Contracts and NASA Adjusted   =   .85
22. S.A., Contract Awards, and S.A., Military Payroll   =   −.05
23. S.A., Contract Awards, and S.A., Civilian Payroll   =   .19
24. S.A., Civilian Payroll, and S.A., Military Payroll   =   .47
25. DoD, Civilian Employment, and DoD, Military Employment   =   .62
26. NASA Prime or Adjusted Contracts and all measures of
    Defense Contracts and Civilian Impact = greater than
    .25 and less than .45
27. All 1967 figures correlated with same variable for 1968   >   .80

except

28. S.A., Contract Awards, 1967 with 1968   =   .66

Table 3.1 (cont.)

Across Years: 1961-62 with 1967-68

29. *S.A.*, Contract Awards	=	.43
30. *S.A.*, Civilian Payroll	=	.89
31. *S.A.*, Military Payroll	=	.83
32. NASA Prime Contract	=	.58
33. NASA Adjusted Contracts (1962-63 and 1967-68)	=	.41
34. ISB, Defense Impact (1963 and 1966)	=	.60

from past wars and are not related to current military acquisitions.

The most important relationships among all these measures are shown in table 3.1. Unfortunately there are a great many figures to assimilate, even after some unimportant relationships have been omitted. But careful attention to certain patterns here will allow us to reduce the quantity of information we must carry over into the most interesting stage, that of relating spending to voting. Items in the table are referred to by number in parentheses to facilitate discussion.

1. Some of these spending patterns are remarkably stable over time, notably the information reported by the *Statistical Abstract* for the Defense Department's military and civilian payroll (items No. 11, 27, 30, 31). These expenditures, largely for the maintenance and operation of military installations, changed little from year to year or even between 1961-62 and 1967-68. This stability, providing a continuing interdependence between the bases and the larger civilian economy of the state, may have substantial political importance. More immediately, the only slight changes from one year to the next mean that we can confidently combine 1961 and 1962 (or 1967 and 1968) expenditures into a single measure for relating spending to Senate behavior in the appropriate congresses.

2. Other spending patterns, particularly DoD and NASA contract awards, are much less stable. Even from one year to the next, DoD prime contract awards correlate only in the .60's with each other (12, 28) and just .43 across the six-year span (29). Most states, therefore, are less likely to be bound in such a continuing way to contracts for industrial procurements of military goods than to the supply of military bases. This is another good reason to expect that the political impact of the two may be different. On the one hand, the risks and uncertainties of military contracting may encourage special political activity by legislators to obtain or hold defense contracts; on the other, the substantial shifts in location of contracting may deprive contractors of the kind of secure basis that may be necessary for continuing political effect, especially on senators who must face the electorate only once every six years. Particular aspects of the shift in contracting also deserve attention, but that will best await a state-by-state listing of defense dependence.

NASA prime contracts also change location substantially (13, 32). Overall the shift is not quite so great as for DoD contract awards, but it is much more so than for DoD direct employment. When the NASA totals are adjusted for subcontracting patterns, the shift is fully as great as for DoD prime contracts (33). Comparable data attempting to adjust DoD contracts over time are not available, though the shifts in defense impact measured by the Independent Study Board, as implied by the correlation of only .60 for two points just three years apart, indicates similar instability (34).

3. Of the three *Statistical Abstract* measures of defense impact (DoD contracts, military payroll, and civilian payroll) contracts are almost completely indepen-

dent of direct payrolls: the correlations range from just .20 to a slight negative correlation (6, 7, 22, 23). Moreover, even civilian and military payrolls are only moderately correlated with each other (8, 24). The distribution of contracts is thus very different from that of military bases, the former, as we shall see, tending to favor some of the industrial states and the latter especially heavy in Alaska, Hawaii, the South, and the Southwest. This reinforces our original decision to keep the measures separate rather than agglomerated into a single index of total "defense impact."[7]

4. Of the various attempts to vary or improve on the *Statistical Abstract* reports, the differences are very slight when they deal with direct employment. The Bolton estimates (2, 3) correlate highly with the appropriate *Statistical Abstract* figures, as do the DoD data for military employment, even though they deal with numbers of workers as a percentage of the labor force rather than payroll as a percentage of income (15). With such high correlations, in the mid-to-high .80s, it matters little which set of figures is used; since only the *Statistical Abstract* data are available for the entire time span we will henceforth use them and discard the others.[8]

5. All the measures concerned in whole or in part with the impact of DoD procurement contracts differ in substantial ways both from the original data on prime contracts (1, 14, 19) and from each other (4, 16-18), with correlations no higher than the middle .60's. (Measures of *total* military impact, including bases and military payroll, obviously are not highly correlated with contract awards, and the coefficients are omitted from the table.) Also, states' rankings according to the impact of NASA contracts change rather substantially when ad-

justed for subcontracts (9, 21), especially in 1961-62 (though the latter is contaminated by the inability to match years precisely).

It seems clear that the prime contracting data do need adjustment for subcontracting in order to provide a fully adequate measure of the impact on states' econo- mies. For DoD impact, however, the major differences among the various attempts to measure subcontracting, coupled with the fact that they necessarily use very in- complete data and rough estimating procedures, make it impossible to select any one as satisfactory for sole use hereafter.

6. NASA's impact on states differs substantially from that of DoD spending (9, 10, 26). Hence the space agen- cy's activities must not be lumped indiscriminately into a "military-industrial complex" when imputing eco- nomic or political effects.

In tables 3.2 and 3.3 are listed states' incomes from various types of defense and aerospace spending, first for 1961-62 and then for 1967-68. The defense or space impact is always expressed as a proportion of the states' total personal income. In accordance with the above discussion, the data for NASA are given as adjusted for subcontracting, and the DoD civilian and military pay- roll figures are as reported in the *Statistical Abstract.* Also, for continuity over time DoD prime contract to- tals are given rather than the figures from any of the efforts to estimate the distribution of subcontracts. To preview the findings of the next section, neither prime contracts nor the revisions for DoD contracting pro- duce high correlations with Senate voting, and it mat- ters little which we look at. The percentages are rounded here, but exact rankings were used in the later correlations.

Table 3.2. States Ranked by Various Types of Defense and NASA Expenditures as Percentage of Personal Income, 1961-62

DoD Prime Contracts		DoD Civilian Payroll		DoD Military Payroll		NASA Contracts, Adjusted	
Utah	18	Hawaii	8	Alaska	20	California	.4
Connecticut	15	Alaska	7	Hawaii	9	Minnesota	.4
California	12	Virginia	6	New Mexico	5	Arizona	.2
Alaska	12	Utah	6	S. Carolina	5	Florida	.2
Colorado	12	N. Hampshire	5	Georgia	4	Connecticut	.2
Washington	11	Alabama	4	Mississippi	4	Iowa	.2
Kansas	10	New Mexico	4	Virginia	4	Maryland	.2
Massachusetts	9	Oklahoma	3	Nevada	4	Massachusetts	.2
Arizona	7	Georgia	3	N. Dakota	4	Pennsylvania	.1
Maryland	6	Maryland	3	N. Carolina	4	New York	.09
N. Hampshire	6	Rhode Island	3	Texas	4	Nevada	.08
Virginia	6	S. Carolina	3	Kentucky	4	Louisiana	.07
New Jersey	6	Washington	2	Maine	4	Colorado	.07
N. Dakota	6	Colorado	2	Colorado	3	Utah	.04
Florida	6	Nevada	2	Washington	3	New Jersey	.03
Texas	5	California	2	Kansas	3	Michigan	.03
New York	5	Texas	2	Oklahoma	3	Ohio	.02
S. Dakota	5	Pennsylvania	2	N. Hampshire	3	Illinois	.02
Georgia	5	Arizona	2	Montana	3	Oklahoma	.02
Maine	5	Kentucky	1	Arizona	3	Missouri	.01
Indiana	5	Florida	1	Maryland	3	Wisconsin	.01
Montana	5	Mississippi	1	Delaware	3	Texas	.01
Ohio	4	Massachusetts	1	Nebraska	2	W. Virginia	.01
Missouri	4	Ohio	1	Florida	2	Rhode Island	.01
Delaware	4	New Jersey	1	Arkansas	2	Alabama	.01
Louisiana	4	Arkansas	.9	Louisiana	2	Indiana	.01
New Mexico	4	Missouri	.9	Wyoming	2	Oregon	.01
Michigan	4	Nebraska	.8	Idaho	2	Kansas	.009
Pennsylvania	3	S. Dakota	.8	Alabama	2	Tennessee	.006
N. Carolina	3	N. Carolina	.8	California	2	N. Hampshire	.005
Minnesota	3	Louisiana	.7	S. Dakota	2	Vermont	.005
Mississippi	3	Kansas	.7	Rhode Island	1	Maine	.005
Wyoming	3	Tennessee	.7	Tennessee	1	N. Carolina	.005
Oklahoma	3	New York	.6	Missouri	1	Arkansas	.004
Tennessee	3	Indiana	.6	Massachusetts	1	Washington	.003
Wisconsin	3	Illinois	.6	New Jersey	1	Virginia	.001

Table 3.2 (cont.)

DoD Prime Contracts		DoD Civilian Payroll		DoD Military Payroll		NASA Contracts, Adjusted	
Alabama	3	Maine	.6	Utah	.9	Georgia	.001
Iowa	3	N. Dakota	.6	Illinois	.7	New Mexico	.001
Arkansas	3	Oregon	.5	Oregon	.5	Mississippi	.000
W. Virginia	2	Montana	.5	Michigan	.5	S. Carolina	.000
Vermont	2	Wyoming	.5	Ohio	.4	Alaska	.000
Rhode Island	2	Delaware	.5	Indiana	.4	Delaware	.000
Hawaii	2	Michigan	.4	New York	.3	Hawaii	.000
Illinois	2	Idaho	.2	Minnesota	.3	Idaho	.000
Nebraska	2	Connecticut	.2	Vermont	.3	Kentucky	.000
Idaho	2	W. Virginia	.2	Connecticut	.3	Montana	.000
S. Carolina	2	Minnesota	.1	Pennsylvania	.2	Nebraska	.000
Oregon	1	Wisconsin	.1	Wisconsin	.2	N. Dakota	.000
Kentucky	1	Vermont	.1	Iowa	.2	S. Dakota	.000
Nevada	1	Iowa	.1	W. Virginia	.1	Wyoming	.000

Looking first at the 1961-62 distribution of defense and aerospace expenditures, the payroll figures of course favor those states in the West and the South that have many or large military bases. Alaska and Hawaii lead both, and six of the top eleven states in the military payroll list are from the one-time Confederacy. For DoD contracts, however, the picture is very different. Alaska is but fourth, Hawaii falls near the bottom, and the first southern state is Virginia, number twelve. Connecticut and California, wealthy industrial states, are in the second and third positions. Since in the nation overall the value of contract awards is typically twice as great as that for military and DoD civilian payroll combined, the effect of defense spending was generally to reinforce existing income disparities among the states, with the rich industrial areas receiving more than their income "share"—doubtless because that is where most of the major manufacturing centers capable of producing sophisticated arms and equipment were to be found. The distribution of bases could only partially modify the effects of contracting in the overall picture. A correlation of states' per capita income with the shares of that income derived from military spending shows a very mild positive correlation, with 4 percent of the variance accounted for. Certainly overall the South gained no economic advantage from high levels of defense spending. Much the same is true for NASA spending, with the major exceptions being Florida with Cape Canaveral (Kennedy), and to a lesser degree the state of Louisiana.

By 1967-68, however, a mere six years later, the situation had changed remarkably. Alaska and Hawaii still top the lists for civilian and military payroll (combined), but below them the importance of the South is strengthened. Of the eleven southern states, seven are in the top half of the DoD civilian payroll rankings and nine in the top half on military payroll (as compared with six and eight, respectively, in 1961-62). Much more important, however, is the rise of the southern states in the contracts list. Eight of the former Confederate states show up in the top half there, and the industrialized states of California, Colorado, and Washington evidence substantial decline, though Missouri gained greatly. Combining payroll and contracts, three southern states—Texas, Georgia, and Virginia—now appear among the top eight. The correlation of total defense impact with per capita personal income, though still a positive one, becomes very slight, explaining only 2 percent of the variation. Under the Johnson administration defense spending became an agent of redistribution of income in favor of some of the poorer areas of the country, especially the South, and most particularly Texas. Georgia, with Richard Russell as chairman of the Senate Armed Services

Table 3.3. States Ranked by Various Types of Defense and NASA Expenditures as Percentage of Personal Income, 1967-68

DoD Prime Contracts		DoD Civilian Payroll		DoD Military Payroll		NASA Contracts, Adjusted	
Connecticut	18	Utah	8	Alaska	16	Alabama	3
Missouri	13	Hawaii	7	Hawaii	8	Louisiana	3
Texas	13	Virginia	5	S. Carolina	5	California	2
Alaska	9	Alaska	5	Georgia	5	Florida	2
California	9	Oklahoma	4	N. Dakota	5	Maryland	1
Georgia	9	New Mexico	4	Virginia	4	Texas	.8
Vermont	9	N. Hampshire	3	Colorado	4	New York	.6
Massachusetts	8	Maryland	3	Kentucky	4	Colorado	.6
N. Hampshire	8	Alabama	3	N. Carolina	4	Arizona	.5
Indiana	6	Georgia	3	New Mexico	4	Massachusetts	.5
Maryland	6	S. Carolina	2	Texas	3	New Mexico	.5
Louisiana	6	Rhode Island	2	Oklahoma	3	Wisconsin	.5
Arizona	6	Colorado	2	Arizona	3	Minnesota	.5
Utah	6	California	2	Mississippi	3	Vermont	.4
Tennessee	6	Washington	2	Montana	3	Connecticut	.4
Minnesota	6	Texas	2	Delaware	3	Delaware	.4
Mississippi	5	Kentucky	2	Washington	2	New Jersey	.3
Rhode Island	5	Arizona	1	Maryland	2	Virginia	.3
Virginia	5	Pennsylvania	1	Nevada	2	N. Hampshire	.3
Washington	5	Florida	1	Alabama	2	Mississippi	.3
Florida	5	Nevada	1	Wyoming	2	Pennsylvania	.2
Kansas	5	Missouri	1	Kansas	2	Maine	.2
New York	5	Mississippi	1	Florida	2	Washington	.2

Ohio	5	Ohio	1	Louisiana	2	Hawaii	.2
Alabama	5	Massachusetts	1	Maine	2	Ohio	.1
New Jersey	5	New Jersey	.9	S. Dakota	2	Iowa	.1
Pennsylvania	5	Indiana	.8	Nebraska	2	Michigan	.1
Colorado	4	Arkansas	.8	California	2	Missouri	.08
N. Carolina	4	N. Carolina	.8	Arkansas	1	Utah	.08
New Mexico	3	Louisiana	.7	Missouri	1	Indiana	.04
Hawaii	3	N. Dakota	.6	Rhode Island	1	Oklahoma	.04
W. Virginia	3	Kansas	.6	N. Hampshire	1	Illinois	.04
Michigan	3	Nebraska	.6	Idaho	1	Tennessee	.04
Iowa	3	Delaware	.6	New Jersey	1	Georgia	.04
Arkansas	3	Illinois	.6	Utah	1	Rhode Island	.04
Wisconsin	3	Maine	.5	Tennessee	1	Alaska	.03
S. Carolina	3	Tennessee	.5	Massachusetts	.7	Kansas	.03
Maine	3	S. Dakota	.5	Illinois	.7	Nevada	.03
Wyoming	3	Montana	.5	Ohio	.4	W. Virginia	.03
Montana	3	Oregon	.4	Michigan	.3	Oregon	.02
Nebraska	3	Wyoming	.4	Oregon	.3	N.Carolina	.02
Delaware	2	New York	.4	Indiana	.3	Kentucky	.01
Illinois	2	Michigan	.4	New York	.2	Wyoming	.01
Oklahoma	2	Connecticut	.3	Minnesota	.2	S. Dakota	.009
N. Dakota	2	W. Virginia	.2	Pennsylvania	.2	Nebraska	.009
Oregon	2	Idaho	.2	Connecticut	.2	Montana	.009
Nevada	1	Minnesota	.2	Wisconsin	.1	Arkansas	.007
S. Dakota	1	Wisconsin	.1	Iowa	.1	N. Dakota	.004
Kentucky	1	Vermont	.09	Vermont	.09	S. Carolina	.004
Idaho	.8	Iowa	.07	W. Virginia	.07	Idaho	.004

Average state income from defense (total across first three columns) is 8.9 percent.

Committee, did well too, in the fifth position overall. This is also true for the distribution of the much smaller total of NASA contracts; the South has but six states in the top half of that list, but four of them are in the first half-dozen ranks. The average (mean) state received 8.9 percent of its income from all defense spending; for comparison the defense income of any particular state can be obtained by summing its figures in the first three columns.

Some Realities about Defense Politics

Tables 3.4 and 3.5 show the rank-order correlations between the various measures of defense and aerospace spending and the several scales for senators' voting. Positive correlations indicate the expected relationship; i.e., rank numbers zero and one from tables 2.1 and 2.4, from the "conservative" or "hawkish" ends, are related to high levels of defense spending. Correlations below .10 are omitted as too low to be of interest and to focus attention on the interestingly high relationships. All correlations of .12 or higher are statistically significant at the .05 level; of .17 or higher at the .01 level, and of .22 or above at the .001 level.[9] The first set of columns shows the simple correlations; the last two control them for two seemingly important political influences—region of the country (former Confederacy vs. all other states) and political party of the senator.

Looking first at the relationships in the 87th Congress, it is striking how little our expectations are borne out. DoD prime contract awards show absolutely *no* relationship to *any* of the three scales of Senate voting, not even when partial correlations are used to control for the confounding effects of regional influences and party

Table 3.4. Rank-order Correlations (Tau) between Senate Voting and Defense-related Expenditures by State, 1961-62

	Uncontrolled			Controlled for Region			Controlled for Party		
	Gen. Def.	Mil. Asst.	Aero-space	Gen. Def.	Mil. Asst.	Aero-space	Gen. Def.	Mil. Asst.	Aero-space
Contract Awards, 61-62 *Stat. Abst.*									
Procurement, 61-62 Bolton		.19			.17			.20	
DoD Civil. Payroll, 61-62 *Stat. Abst.*		-.12					.13	-.13	
DoD Mil. Payroll, 61-62 *Stat. Abst.*	.14	-.22		.11	-.19		.21	-.24	
Total Def. Impact, 63 ISB			.17		.10	.17			.15
NASA Prime, 61-62					.12				
NASA Adjusted, 62-June 63		.18	.12		.19	.12		.19	.11

loyalty. This failure cannot be attributed solely to inadequacy of prime contract data for getting at the "real" defense impact, since neither the Bolton procurement estimates nor the Independent Study Board's estimates of total defense impact improve the situation with the most important scale, that on general defense, East-West relations, and COMSAT matters.

In fact, the only defense spending measure that shows a strong relationship at all with that scale is military payroll, where the effect is in the expected direction.[10] The association is strongest when state-by-state spending is controlled for party; when that happens a very mild correlation between spending and hawkishness also appears for the Defense Department's civilian payroll. Republicans tend to be more "hawkish" than Democrats, but within both parties hawkishness on the general defense scale is related to spending. The relationship nevertheless is with direct employment, largely to man military installations, and not with contract procurement. And the more "civilian" the installations (the less the share of the military payroll), the less the impact. Furthermore, the correlation weakens somewhat when one controls for region. Southerners, with the bases, benefit most from the direct military payroll and their senators are also more hawkish, on the average, than are northerners.

On the rather narrow economic interest expectations, NASA expenditures are related, still to a very modest degree, with approval of aerospace programs in the Senate. That appears, nevertheless, only when the NASA prime contract data are adjusted for subcontracting. Surprisingly, the Independent Study Board's defense impact estimates, though not related to the general defense scale, show a stronger correlation with aerospace voting than to the NASA totals. Possibly this is slightly less surprising, though still curious, when one recalls that the ten-item "aerospace" scale did include two civil defense roll calls and an item concerning foreign aid to communist countries.

The strongest relationships in the entire table are with the scale of votes on military assistance, but many of the correlations are not in the expected direction. Bolton's data on military procurement show a positive relation to enthusiasm for foreign military aid, as do the adjusted figures for NASA contracts. But direct employment of civilians, and especially military personnel, is associated with opposition to such aid, and this holds despite the controls. It is worth remembering that in the early 1960s most "liberals" still favored foreign assistance, even of the military variety, and that the military assistance scale showed a moderate negative correlation with the general defense scale. The "doves" on general foreign policy issues often approved of military aid. Further discussion of this and other points can best be held until the 90th Congress data have been examined.

Some of our previously unfulfilled expectations do a little bit better in table 3.5. The relationships between the general defense and East-West relations scale and DoD civilian and military payrolls are stronger than before. Again they hold even when party and region are controlled, and again the relationship is strongest with the military payroll. Clearly *Department of Defense expenditures for military installations go to support and reinforce, if not to promote, a set of hawkish and strongly anticommunist postures in American political life.* This support may very well be inadvertent rather than deliberate, but it does exist. In turn, the Pentagon is supported, and its expenditures promoted, by those

Table 3.5. Rank-order Correlations (Tau) between Senate Voting and Defense-related Expenditures by State, 1967-68

	Uncontrolled				Controlled for Region				Controlled for Party			
	Gen. Def.	NASA	Gun Cont.	Arms Sales	Gen. Def.	NASA	Gun Cont.	Arms Sales	Gen. Def.	NASA	Gun Cont.	Arms Sales
Contract Awards, 67-68 *Stat. Abst.*			-.15				-.20				-.12	
DoD Civil. Payroll, 67-68 *Stat. Abst.*	.16	.10			.11				.20	.11		.11
Def.-Generated Civil Empl., 67-68 DoD		.11	-.17			.11	-.20			.13	-.13	.14
DoD Mil. Payroll, 67-68 *Stat. Abst.*	.26	.14	.15		.19	.12	.11		.28	.15	.19	
Total Defense Impact, 66 ISB		.10	-.18			.10	-.20	.10		.11	-.16	.13
Priority One Def. Impact, 66 Steger			-.17				-.19				-.16	
NASA Prime, 67-68		.13	-.25			.13	-.27			.14	-.25	.11
NASA Adjusted, 67-68		.16	-.24	.13		.16	-.26	.13		.17	-.24	.14

voters and political leaders. And more than just military spending levels are at stake. Many of the defense-related roll calls on which the doves or liberals were defeated in the 90th Congress concerned highly controversial matters, especially regarding military expenditures. There they were opposed by a broad coalition of conservatives and moderates. But many of the votes that were just won by the liberals or on which they were narrowly defeated were not at all extreme, at least by this liberal's view: in 1967-68, the Consular Convention with the Soviet Union; a 3 percent reduction in military procurement; a $20 million authorization for the Arms Control and Disarmament Agency; withholding funds for the ABM until the cost and practicality of the system were known; reform of Selective Service procedures. These were mild measures, offering very tentative steps toward better East-West relations or minor restrictions on the military establishment. The same applies to some of the 87th Congress votes for example, support for the United Nations Congo operation and authority for the Arms Control and Disarmament Agency to do research on disarmament! Surely some such steps and more, reciprocated by the communists to be sure, will have to be taken if the cold war is ever to end.

The correlations are not, it must be noted, astonishingly high—nor did our original expectations, which recognized the data handicap of using heterogeneous states as the unit of analysis and identified the probable sources of many deviations from the simplest expectations, anticipate very high correlations. The distribution by state of defense spending is surely only a partial explanation of senators' votes even on DoD appropriations and authorizations and less on a wider spectrum of issues. But the civilian and military payroll of DoD are not trivial influences either.[11] Furthermore, there are clear signs of the emergence of a stronger relation between defense payrolls and senators' voting behavior, just as in the last chapter we found the emergence, from latent to manifest, of an alignment on defense matters and East-West relations effective across a rather wide set of issues. Finally, there are indications that some senators take their rewards not in defense spending for their districts but in leverage over other political matters. More on that later.

Let us look briefly at the minor defense-related scales. Among the simple correlations, and with a control for region, no class of military spending is much related to votes for or against promoting arms sales abroad. Some effects, however, are masked by party differences, and Republican senators are somewhat more likely to favor foreign arms sales than are the Democrats. With party membership controlled, small but sometimes statistically significant relationships emerge for several of the employment or defense impact measures. But no such relationships appear for either prime contract awards or for the military payroll, and since the other correlations are low probably not too much should be made of them.

The gun control situation is more puzzling. Most of the defense spending measures—and those for NASA too—show moderately high correlations, yet not in the expected direction. As anticipated, states with many soldiers (high military payroll contribution to personal income) tend to have senators who look with disfavor on gun control legislation. Yet all the other measures (except DoD civilian payroll, with no notable correlation either way) show that senators from states with heavy defense impact are more likely to approve of restrictions

on the importation and sale of firearms. This is true
when region and party are controlled, though the latter
weakens it somewhat. Republicans generally like gun
control less than do Democrats; when one looks within
each party instead of lumping them together the positive correlation with military payroll is strengthened,
and the negative correlations for the other measures become a bit lower. Nevertheless they remain notable and
statistically significant. But much of this fades further
when one controls for states' urbanization, on the principle that the urban, industrialized states generally need
and approve gun controls more than do the rural states
where hunting is widespread and the spirit of individual
self-reliance is possibly more active.[12] Control for urbanization reduces all the negative correlations substantially. For NASA they are cut in half, and none of the
others remain greater than $-.12$. Thus heavy defense
spending does not very notably encourage a state's representatives to favor gun control legislation, but neither
does it predispose them against it, as we originally sup-
posed. Perhaps the reason is in the element of gun control that is directed toward the importation of foreign
firearms. Some domestic arms-makers have benefited
handsomely from those provisions and expected to do
so when the legislation was being considered. Then too,
some gun manufacturers felt that unless existing abuses
were curbed somewhat the alternative might be even
stricter controls. The relatively moderate proposals of
Connecticut's Senator Dodd, for example, seem to have
been acceptable to most of the arms companies and
military contractors in his own defense-rich state.

Contracts and Bases

The relation of the state-by-state distribution of NASA

spending to senators' approval of NASA's activities is about as expected, though not especially high. It holds with the usual controls and is appreciably higher than the similar relationship found in 1961-62. It also is stronger, by .03, for the adjusted contract totals than for the prime contract figures alone. This difference nevertheless is not great, and it leads to a question about the major anticipated but absent correlation in these tables—between defense contract awards and voting on the general defense scale. Any temptation to attribute its absence to poor data quality on contracts must be rejected. None of the three efforts to make some estimate of subcontracting impacts—the DoD figures on defense-generated civil employment and the defense impact estimates of Steger and of the Independent Study Board—help at all, except for a mild correlation for the first when party is controlled. And if the improvement to be expected from good subcontracting figures is no better than what is evidenced in the NASA correlations, that is hardly the answer. It seems unlikely that access to classified data on subcontracting would show any appreciable difference.

The data can be looked at still more intensively and from other perspectives. For one, it might seem plausible that the urban industrial states are inherently more liberal or dovish on defense and foreign policy matters. True, but even with urbanization controlled, partial correlations worthy of note still fail to appear. Or it might seem likely that the connection between defense contracts and voting is a delayed one; that is, contract inputs come later than "pro-defense" votes in the Senate. That could happen either because the votes themselves authorize the specific contracts or because grateful officials in the executive branch reward cooperating senators. But this is not the case either. Correlation of the 1961-62 general defense scale with 1963 prime contracts and also with the 1967 and 1968 prime contracts made no difference, though it did produce some mild (.11-.15) correlations between 1963 contracts and the other two scales. Another way of asking the same question says that the most hawkish senators in the 87th Congress should see their states rise on the contract rankings by 1967-68, and the most dovish senators' states should suffer a decline in the rankings. Yet a check of the figures, comparing the most hawkish quintile with the most dovish 20 percent (actually 21 senators in each case because of ties) shows no notable difference between them in the number of states that declined or rose in the rankings.

Another possible refinement reads as follows: "It is a mistake to look at the entire range of senators on the defense scale, because to most observers, including constituents, lobbyists, and Pentagon officials, the great middle is an undifferentiated mass among which the observers cannot make fine distinctions about hawkishness or dovishness, and hence cannot effectively apply political pressures. Only by concentrating on the extremes of the scales could one expect to find the basic expectation clearly confirmed." This view is strengthened by some of our initial propositions, which suggested that though hawks from high-defense-impact states would be especially common, there would be enough other influences generating hawkishness in middle- and even low-impact states to depress seriously the overall correlation. The relation should be most strongly manifested only in the near absence of doves from high-impact states, and, where such doves did nevertheless exist, only special circumstances could explain their behavior. To check this possibility, I took the 20 most hawkish senators in 1967-68 (when the existence of a defense scale was more manifest than in 1961-62) and compared them with the 20 most dovish, asking whether they were from the 20 percent of states with the greatest contract impact, from the 20 percent with the least impact, or from the middle 60 percent. By pure chance alone, if defense spending had no relation to voting on the defense scale, one would expect to find in each group 4 senators from each of the high- and low-defense-impact states and 12 from the middle. The actual results are seen in table 3.6.

Table 3.6. Defense Contract Disbursement and Voting Behavior among the Most Hawkish and Dovish Senators, 90th Congress

	From 10 states with greatest contract impact	From middle 30 states	From 10 states with least contract impact
Strongest hawks*	6	11	4
Strongest doves	3	12	5
Would expect by chance	4	12	4

*Actually there are 21 strong hawks, with 16 tied at the first rank and 5 at the next. Of the 5, 1 is from a high-impact state, 2 are from low-impact states, and 2 are from the middle ones.

The results are slightly, but only slightly (and not statistically significantly), in the direction of the revised prediction. Strong hawks are a bit more common from high-impact states than would happen by chance, and doves are slightly less common from high-impact states and more frequent in low-impact states. And special circumstances do operate in the case of some doves. Alaska is in a unique position vis-à-vis the defense effort. Substantial installations must be maintained there for continental air defense; these installations are not heavily manned but are costly to maintain. Moreover, distance and high transportation costs make it expensive to bring goods into the state, so many goods and services for the bases must be obtained within Alaska. Yet there is little of the conventional aerospace kind of industry found in Washington or California. Thus Senator Gruening, one of the three exceptions in the lower left box, was virtually immune from the most direct "punishment" for his vociferously dovish position—his state could not be deprived of many contracts without appreciably hurting the defense of the country. Similar if less dramatic cases of unique contracting capability by a particular state or region may help explain other, smaller, deviations from the expectation. Another of the three high-impact doves was Senator Yarborough of Texas. Senator Yarborough's relations with the Johnson administration were always strained, but for obvious reasons his state, if not Yarborough himself, was forgiven for his dovish behavior. No such explanation is readily available for the third exception, Senator Hartke of Indiana.

Nevertheless, if heavy defense contracting produces or reinforces a hawkish viewpoint among a senator's constituents, eventually he will be in electoral trouble. Senator Gruening was one of only three incumbent Democratic senators to be defeated in 1968, and Senator Yarborough fell in a May 1970 primary. Their states also are high on the military employment rankings.

One dovish senator from a high-impact state may enjoy a somewhat protected position if his colleague is sufficiently solicitous to the needs of local defense industry. A single senator may be able to nurture and conduct specific proposals through the legislative process, and crude retaliation against the dove, by denying contracts to his state, would punish the hawk also. Senator Yarborough thus benefited indirectly from Senator Tower's very hawkish position. Two other high-impact states, California and Connecticut, may now be moving to fit this situation though they did not in the 90th Congress. Each has a very hawkish senator, Murphy and Dodd respectively, the latter unique among northern Democrats. The new senator from California, Alan Cranston, is known as an outspoken dove, and Senator Ribicoff, formerly near the middle, has been moving "left" since his 1968 speech in Chicago.

This evidence, plus the mild party-controlled correlation of hawkishness with defense-generated civil employment, plus the original expectation that defense contractors, unlike the local suppliers of military bases, would attempt to exert political influence on senators from other states as well as their own, leaves us with an impression of some relation between defense contracting and senators' votes on defense-related issues. It may be stronger in the House, with smaller and more homogeneous constituencies. Furthermore, lobbying for contracts concentrates on particular weapons systems offering narrow and limited demands for senators' support, whereas senators whose home states are economically dependent upon military bases often must support overall high levels of military spending in order to get benefit for their constituents. Certainly senators with defense industries at home do look out for them. California, Colorado, and Washington represent three of the centers of the aerospace industry; analysis of individual votes by their senators on 20 specifically aerospace questions in the 90th Congress shows exactly one vote against the aerospace industry out of a possible 120.

But this kind of direct concern for constituents' industry was to be expected in the operation of any political system. No matter how we look at the data, however, the relation between DoD payrolls and voting behavior, on a constellation of domestic, defense, and foreign policy issues, just does not exist in comparable degree for contracting. Perhaps military bases, because they are stable and enduring, exert a political influence on Capitol Hill that here-today, gone-tomorrow government contracts cannot. In any case, readers who otherwise found their prejudices against the "military-industrial complex" being confirmed by this study must absorb this finding too. The industrial part—that is, the big manufacturing establishments—does not reinforce the hawkish or uncompromisingly anticommunist forces in this country in any strong, simple, or direct way. Military spending on particular weapons systems may well be promoted by industry, and, insofar as those weapons are redundant for national defense, the spending is wasteful. But the political effect of that spending is not the same as that of money spent to maintain a large army of many men, with bases scattered freely across the country.[13]

I said earlier that some modification of extreme hawkish attitudes was necessary for the long-run wellbeing of this nation. Some who agree may still continue to oppose all kinds of military spending. They might do well, nevertheless, to differentiate that spending according to its political effects, by the limited evidence of this study. A smaller army, navy, and air force, even provided with expensive and heavily automated equipment and with no reduction in overall spending, might less strongly reinforce (note that I do not attribute a simple causal effect) hawkish elements in the American political system. Perhaps that is an argument, to be weighed against others, for abandon-

ing Selective Service in favor of a smaller volunteer
army.

Defense and Civil Rights

We may close by looking at the behavior of southern
senators on defense issues and of northerners on civil
rights matters. In his important study Samuel Huntington showed conclusively that in the 1950s Democrats
were *more* likely to be pro-defense than were Republicans. This was true for senators from all of the major
sections of the country. And southern Democrats were
not notably more pro-defense than were their northern
brethren.[14] But from the data of this book, surely in
the 1960s northern Democrats were on the whole much
less hawkish than were either Republicans or southern
Democrats. In the two congresses from the 1960s
looked at here, only a few of the strongest doves were
Republicans or southerners, and with only a single exception (and that only in the 90th Congress) all the 20
to 25 most hawkish senators were Republicans or southern Democrats. As many as 9 of the 22 senators from
the old Confederacy were among the 16 legislators with
zero scores.

The reversal between northern Democrats and Republicans can in large part be explained by the politics of opposition. "Preparedness" was a good vote-
getting issue throughout the 1950s and most of the
1960s. The outs could register a popular complaint
that not enough was being done for defense; the ins
usually felt obliged to defend their President and his
administration. But that does not explain the southerners' behavior, particularly so since their hawkishness
in 1961-62 was not accompanied by substantial eco-

nomic benefits to their constituents from defense spending. Rather, their states lost money overall. Perhaps their relative gains from military bases (though not from contracts or total defense spending) made the difference, in light of our previous suggestions about why military installations may be more politically important than are big contractors. Or very possibly southerners became hawkish in anticipation of more economic benefits. Certainly by 1967-68 the South had its material reward for supporting defense spending. Also, our finding of an association between votes on DoD expenditures and other defense and East-West relations issues would of course support the hypothesis of an ideological rather than purely logrolling basis for defense support.

But another and rather subtle argument has also been offered, one that we can neither reject nor confirm. It has been said that southern congressmen advocate heavy defense spending, despite their general predisposition toward fiscal conservatism, as a means of maintaining their own power within the legislature. Southerners have long held strategic positions on congressional committees dealing with the armed services; Senators Russell and Stennis and Representative Rivers are merely among the more dramatic examples. Their power to ease or block particular military procurement, it is said, gives them a great potential source of power over northern Congressmen. Representatives from states or districts with a heavy defense impact will, it is alleged, fear to antagonize the powerful southerners on the key defense committees. Hence they will often, perhaps in quiet ways, mute their own support of civil rights measures and also support the southerners on questions of organizing the Senate. Thus a high level of defense spending would be a keystone in maintaining the traditional southern political and social order, and continued heavy military expenditures would have the effect, whether or not intended, of helping to maintain racial discrimination.[15]

Certainly southern senators usually oppose civil rights legislation, and there is a relationship between voting on defense and on civil rights. As a preliminary effort we performed the scaling operations on several large sets of roll calls, including a range of issues covering all the votes, foreign and domestic, coming before the Senate in both the 87th and 90th congresses, using a somewhat looser criterion of association than was finally employed with the defense-related roll calls alone.[16] In the 87th Congress the defense and East-West relations votes were rather scattered, falling into no consistent association with any particular scale for other kinds of issues. In the 90th Congress, however, they tended strongly to group in a much larger scale that was concerned additionally with civil rights and urban affairs. This finding is new as compared with previous efforts to scale congressional roll calls, including those of MacRae and Rieselbach.[17] Further analysis shows that a separate scale of "conservatism" on civil rights and urban affairs roll calls, derived by John McCarthy from the second session of the 90th Congress in the same way as were the other scales of chapter 2, correlates fairly highly (.60) with the general defense and East-West relations scale.[18]

On the relation between civil rights voting and state-by-state defense expenditures we can say the following: Among the 22 southerners, defense spending does not seem to work against acceptance of civil rights. On the contrary, contract awards and the Steger defense impact estimates seem associated with limited *support* for civil rights (significant at the .05 level, unaffected by the usual controls). This suggests that defense contracts go primarily to those southern locations where traditional racial attitudes are being modified, and perhaps they encourage that change. Yet military payroll shows, at the same level, a correlation with opposition to civil rights. Once again, the presence of military installations is associated with—and perhaps produces—a variety of illiberalism. But there is so little variation in civil rights voting among southern senators as to make this interpretation tenuous.

The real question concerns the leverage of certain powerful southern solons over defense-dependent *northerners*. Thus we want to know whether the voting positions of the 78 northern senators are correlated with any of the measures of defense spending, with effects that may be unknown and unintended by many who advocate greater preparedness. In fact, not much relationship is evident. Of all the spending measures, only DoD military payroll as reported in the *Statistical Abstract* and military personnel as reported by the DoD itself show any correlation with opposition to civil rights legislation among nonsoutherners. (The correlations are .14 and .17 respectively; the .05 statistical significance level for 78 senators is .15.)[19] These relationships hold when party and urbanization are controlled, but they are not very high, and nothing appears for other kinds of DoD spending, such as contract awards.

We must be wary of pushing any interpretation very far. There is some relationship, and perhaps it would emerge as a stronger one in earlier years' data. With the focus of civil rights activity now on the North, by the 90th Congress southern senators may no longer have cared enough to exert extraordinary resistance efforts even if they once did so. On the other hand, we must remember the fact that in the 1950s southern legislators were not especially prodefense. The question must be left unsettled here, but it remains worrisome and would repay investigation of congressional behavior at a more detailed and probing level than is possible with roll-call analysis.

4. Alliances and the Price of Primacy

Alliances and Collective Goods

Both the United States and the Soviet Union have tried to ease their defense burdens by acquiring numerous allies. Formal alliances may have many purposes, of which the pursuit of military security is but one. A state may ally with another in order to achieve direct "side-payments" such as territorial concessions or economic assistance. Or a state may seek arms shipments in return for alliance with a larger power, in order better to restrain its own populace instead of for reasons of international politics. Even alliances concluded with external threats in mind may pursue security in either or both of two ways. A state may take on an ally in order to augment its own power, hoping that the new ally's army, when combined with its own forces, will be enough to deter or defeat a prospective opponent. This is perhaps the most common consideration in classic military alliances. In other cases, however—and these seem especially common for the United States in the post-World War II period—the alliance serves to extend an umbrella from the strong power to cover weak allies. In the long run and with more than immediate military considerations in mind, the large state may consider the smaller one important to its security even though the small one's armed forces are weak. Thus the individual Scandinavian or Low Country allies make little increment to NATO military power, but the United States hopes to add its strength to theirs and so deter any communist attack on them.

This chapter was written with the collaboration of Harvey Starr.

Contemporary American alliances, and perhaps also those of Russia, serve either or both of the last two purposes for the great power. America and the Soviet Union are overwhelmingly suppliers, not recipients, of economic and military assistance in their respective alliances (Russia after World War II was a partial exception), so most side-payments are unimportant to them except as they use them to entice smaller states into alliance. What the superpowers seek is to augment their own military strength with that of their allies or to extend the deterrent umbrella. But insofar as the first is important, both superpowers regard most of their alliances as by no means the successes for which they had hoped. The Soviets long have had good reason to question the reliability of their allies in any actual combat with Western troops. Even now one wonders whether Hungarian and Czechoslovak troops—especially if they were not fighting West German forces—would constitute a net gain or a loss to the communist side in a war. Conceivably they could pin down nearly as many Soviet troops to insure their allegiance as they would add to the collective strength.

The Western side may not have many worries of this sort, but United States military and political spokesmen have long complained that their allies provide less than a fair share contribution to the joint defense. A well-known analyst of American foreign policy exemplifies this very widely held view: "Two decades after a war from which they have long since recovered economically, they show few signs of fulfilling the original postwar expectation that they would assume the major burden of their own defense in return for an American guarantee."[1] In all of its multilateral alliances—NATO, SEATO, and the Rio Pact—the United States devotes a much larger proportion of its national income to defense than does *any* other alliance member. This becomes a matter of acrimony, especially at NATO, when members' contributions are reviewed. Apparently the Soviet Union has the same problem, since it devotes a larger proportion of its income to military purposes than does any other member of the Warsaw Pact. And Russians too seem to feel that their allies are freeloading.[2]

The problem is a basic one in alliances among states which, though juridically equal, are very unequal in power. If the bigger power hopes simultaneously to extend its umbrella to protect the smaller ally and to gain the ally's resources as an increment to its own, the two goals come into conflict. For so long as the smaller state is neither coerced by the big one nor offered special incentives, and unless the threat to the small state is very grave indeed—as in actual wartime—the small nation is likely to regard the big country's armed forces as a substitute for its own. The small country will feel able to relax its own efforts because it has obtained great power protection. Thus the big power's success in extending its umbrella works against its other goals of using the alliance to enhance its own military strength. The small power's effort will vary inversely with its confidence in the big power's guarantee and the disparity in size between the two.

This matter of alliance burden-sharing is an aspect of a wider concern among economists with the theory of "public" or "collective" goods, produced by organizations whose function is to advance the common interests of members. The theory, and the evidence for it, are important in any understanding of alliance behavior. It was originally set forth by Paul Samuelson, was further

developed by John Head, and has recently been applied
to alliances and other small groups in important studies
by Mancur Olson and Richard Zeckhauser.[3]

Public, or collective, goods are defined by two properties. One is that of "external economy" or "nonexclusiveness." External economies are benefits that are made
equally available to all members of a group; it is not
possible or economically feasible to exclude non-
purchasers from the benefits. The other is called "non-
rivalness" or "jointness" of supply, meaning that each
individual's consumption leads to no subtraction from
the supply available to others; this implies that the additional or marginal cost to others, if a good is provided
for one, is small or actually zero. These properties are
conceptually distinct, and both are required to identify
a public good. "Private" goods lack both of these properties; "mixed" goods may have some degree of both or
of one and not the other.

At the *national* level, deterrence is virtually a pure
public good. Under most circumstances an attack upon
any one part of the country will meet with as strong a
response as an attack on any part of the nation. It is
neither militarily nor politically feasible to exclude Mis-
sissippi or California from the deterrent umbrella that
covers the entire United States. And once the deterrent
force is provided for any 49 states, inclusion of the 50th
in no way diminishes the security provided to the
others.[4] In a military *alliance,* the collective good aspect
remains very strong but is somewhat mixed. Nonexclusiveness is not necessarily met, because of a nation's will
or ability to regulate the credibility of deterrence. As
Ypersele notes, "It is partly an awareness of this possibility which had led France to start building its own
deterrence."[5] The great power may indeed want to limit

the credibility of its assurances, since they carry a risk of involvement in major war for the sake of allies' objectives that to it may be trivial. Even with resolute will and high capability, the amount of "automatic" security provided by the United States in NATO, for instance, will not be the same for its allies as for itself nor the same for all allies. It will unavoidably be influenced by such factors as geographic proximity and the web of almost intangible political, social, and economic ties between the big power and the states it would protect. Consider, for example, the differences between Canada and Turkey, the least and probably the most exposed to Soviet ground forces, and respectively probably with the strongest and weakest intangible ties to the United States.[6]

Let us nevertheless assume for the moment that deterrence is the only function of a military alliance and that the deterrence provided by any one member for itself becomes entirely a public good for all other members. (Both these assumptions will be relaxed below.) The biggest power will buy a lot of deterrence, much more than any of the small powers would buy for itself given the same tastes and costs the big power has. Because in the alliance they get this deterrence free from the big power, they also will buy much less for themselves than they would do in the absence of an alliance. But how much will they buy? Will smaller nations all make approximately the same proportionate contribution to collective deterrence (but a much smaller proportionate contribution than the big power), less increasingly in proportion to their smallness, or in fact none at all?

Olson and Zeckhauser show, in a technical argument that need not be repeated here, that under reasonable assumptions the larger the nation the more dispropor-

tionate the share of the total military cost it will bear.[7]
In small groups this effect may be mitigated because
freeloaders can be spotted, and their laggardness may be
highlighted by publicity about members' contributions
(for example, the NATO annual review and press re-
leases showing the defense budgets of all members).
Thus the larger the group the greater the proclivity for
"exploitation" of the big members by the small ones.
Some form of group coordination and/or organization
may provide social incentives for smaller states' contributions, but short of actual coercion (and the theory is
meant to apply only to voluntary groups) some disproportionality will remain.

There are even cases where the small powers will provide nothing.

> In some small groups each of the members, or at
> least one of them, will find that his personal gain from
> having the collective good exceeds the total costs of
> providing some amount of that collective good; there
> are members who would be better off if the collective
> good were provided, even if they had to pay the entire
> cost of providing it themselves, than they would be if
> it were not provided.[8]

In such small groups (which Olson labels "privileged")
where it is worthwhile for some member to provide all
the good, the others may indeed not buy any.

But as we have noted, even if the purpose of military
expenditures were only to provide deterrence against
external threats, in a real-world alliance only a portion
of the security obtained for itself by the large country
can be extended to its allies, due to problems of credibility. As said above, the characteristics of a pure public
good are rarely met in situations of alliance, and thus

the smaller countries will have to buy some measure of security for themselves. Because the security provided by the big country is only partly a public good for its allies, the total derived from the big power alone may be less than the ally would have provided if it were not in alliance. Thus the ally is induced to supplement this automatic security. This supplement is also a partial public good, and "the smallest allies will, in turn, benefit from the additional security provided by the medium-sized countries" all along the line. Therefore the smallest countries will incur miniscule (but probably greater than zero) military expenditures.[9] The big countries also get *some* very modest benefit from the supplement provided by the small, and hence everyone's expenditures are reduced by the alliance, but for the small nations not by as much as if deterrence were a pure public good.[10] In a narrow economic evaluation, the big power must ask itself whether the increment to its own security provided by its allies' efforts is greater than the additional cost (for instance, in maintaining foreign air and missile bases or a "tripwire" presence) it incurs because of the alliance.

If we relax the assumption that military expenditures only provide deterrence, other reasons for military spending by small allies become apparent. *Defense* or damage-limiting capability if deterrence fails, even in the purely national setting, is less clearly a public good than is deterrence. There will be questions about which cities to defend with ABMs or, in West Germany, whether to draw the line of battle at the East German border, at the Rhine, or somewhere between. In an alliance these choices are all the harder, and each ally is almost sure to give priority to its own defense at the expense of others.[11] Defense does not meet either criterion (non-rivalness or nonexlusiveness) for a public good. For these reasons few nations will rely entirely on the deterrence and defense forces provided by alliance with a larger power but will supplement as private goods from their own resources whatever is provided by the alliance. Great concern about defense, however, implies a failure of the alliance to provide for its members a high level of confidence in the pact's deterrent capability. In wartime or periods of acute threat, defense may become a "superior good" defined as something for which an amount at least equal to all income increments will be spent.

Furthermore, there will usually be some private returns from military expenditures, benefits that have nothing to do with defense or deterrence. Overseas colonies are explicitly excluded from the NATO umbrella, and colonial powers must themselves provide for their security. Ground forces may be needed to maintain internal security or to restrain rebellious colonies. Some kinds of military research and development offer potential civilian benefits. French and British leaders have often been unwilling to cut themselves off from large areas of technological advance as the price for forgoing certain kinds of weapons development.[12] For the same reasons nations in alliance will be reluctant to specialize too heavily according to the principle of comparative advantage—for instance, supplying infantry but giving up military R & D.[13]

The analytical distinction among the functions of military expenditure, cast in terms of collective and other goods, is important because of the prediction that to the degree the alliance provides a collective good, the smaller allies will lack incentive to raise armed forces of their own and instead will rely largely upon the great power member(s). Thus the failure of smaller NATO and Warsaw allies to match their protectors' proportionate exertions could be explained in very general rather than ad hoc terms. It would be seen as a "normal" consequence of alliance rather than a lack of ideological fervor or, as has sometimes been alleged, a common European preference to shirk burdens and rely upon others wherever possible.

In addition to refuting that kind of pejorative argument, the applicability of the theory would indicate a measure of the success of NATO (or Warsaw) in meeting at least one of the aims of its superpower originator—to extend the umbrella of deterrence so that other nations will have confidence in the will and ability of the superpower to resist or avenge attack on its clients. The *failure of burden-sharing* would thus indicate the *success of deterrence,* and the initial hope of American policy to obtain a high degree of both would be fundamentally contradictory. We will therefore look at the relative applicability of the collective goods theory at different points in time, associating changes with the evolution of strategy and political cooperation. Also, we can compare the NATO and Warsaw Pact experience with that of other multilateral alliances in the post-World War II era. Various alliances are affected in different ways by big power dominance, and the alliances serve a different mix of functions. By comparing the distribution of burdens actually borne with that predicted by the theory of collective goods, we can better understand what these functions are in each case. Where some or all members spend more than predicted, we have three possible explanations to weigh:

1. They lack confidence in the resolve of their protector (the nonexclusiveness criterion is not met) or in his ability.

2. They seek private goods (e.g. internal security, research and development, control of colonies) from military spending.

3. They are coerced by the big power into spending more than they would (the alliance is not a voluntary organization).

Finally, we will be able to make a general assessment of the value of alliances to the United States and of the degree to which, if at all, it is reasonable to hope that a significant reduction in American military burdens will be made possible by our allies' contributions.

Relative Defense Expenditures in NATO

There is no question but what the pattern of proportionate military expenditures within NATO is in the general direction predicted by the theory of collective goods. Olson and Zeckhauser, Ypersele, and Pryor all have tested the relationship between size and proportionate military expenditures (hereafter referred to as the ratio of defense to gross national product, or simply D/GNP), and found positive correlations. Ypersele and Pryor, in fact, each did so for two different points in time (1955 and 1963, and 1956 and 1962 respectively, always finding a positive relation). Olson and Zeckhauser also found the same relationship between nations' size and both their contributions to the United Nations as a proportion of their assessments and their foreign aid to underdeveloped countries as a proportion of national income.[14] These are the only tests with data on international relations, however, and cannot be used to generalize to the entire population of alliances. A systematic study of NATO and of Warsaw over time is needed, as is a study of burden-sharing in other alliances. Also, some methodological refinements to previous studies will lead to new findings.

All the previous studies have used GNP as a measure of size and D/GNP as an indicator of relative effort. Since economic size is really the implicit variable in the theory, GNP seems appropriate rather than area, population, or some other measure. GNP is typically used as a general indicator of a nation's resource base. The measure of effort, D/GNP, is not flawless due to some incomparabilities of the measure between nations, for example the reliance of some countries on conscript armies that may be grossly underpaid compared with civilian wages. In such cases the defense budget will be less than the cost to the country, in productive resources, of the military effort.[15] But in general the distortions are not so serious as to prevent us from using it.

The other methodological point concerns the measure of association. The virtue of a rank-order coefficient like Kendall's Tau is that it does not assume that GNP in U.S. dollars is a perfect interval measure of national size. Also, the relative rankings may be more important than nations' absolute size differences. On the other hand, it may be useful to see the effect of the intervals. The difficulty there is that the product-moment correlation coefficient will be distorted if one nation is markedly larger than the rest. Because the United States and the Soviet Union are economically very much larger than the second-ranking states in their respective alliances, the correlation will mostly reflect the relative standing of the superpower on the GNP and D/GNP lists, virtually ignoring the other states. To prevent that we have transformed all GNP figures to their logarithm (base 10) whenever (and only when) a superpower is included in the calculations. An alternative is simply to omit the great power. In any case, it is important to look at the plot of nations' positions to see whether the correlation coefficient is hiding information, and to understand the deviant cases which are poorly explained by the general relationship. Further methodological discussion is unnecessary here—each procedure complements the others, and we will use them all to extract as much information as possible from the data.

Table 4.1 gives GNP in U.S. dollars for every NATO country in every year for the period 1950-67, and table 4.2 does the same for the D/GNP ratio. Table 4.3 gives each of the four measures of association for each year, allowing us to evaluate changes over time. It also shows the mean or average D/GNP and the standard deviation (a measure of dispersion) for all NATO nations except the United States. The reason for these will be apparent later.

A great deal of information is implicit in the figures in tables 4.1 through 4.3. First, it is evident that ever since the first years of the Korean War NATO has fit, in general outline, the prediction made by the theory of collective goods. Almost all the correlations are statistically significant, and most often at the .01 level. In that respect our data merely confirm, over a longer time span, the findings of earlier studies. The 1950 measures of association all are quite low, but that was before the United States and Canada built up their forces for use in Korea and to meet a newly perceived threat of Soviet aggression elsewhere. Unlike all the other countries, Greece and Turkey show declines in their D/GNP ratios between 1951 and 1952—these small nations of course entered NATO only in 1952. Although previously under American guarantee through the Truman Doctrine, they seem to have responded strongly to the formal assur-

Table 4.1. GNP at Factor Cost in Billion Current $U.S. for All NATO Countries, 1950-67

	1950	51	52	53	54	55	56	57	58	59	60	61	62	63	64	65	66	67
U.S.	262.6	303.9	319.2	334.6	332.8	364.3	387.1	410.5	414.9	449.5	466.2	481.7	517.0	544.5	585.4	636.4	698.8	737.4
W. Ger.	20.1	24.3	27.7	29.5	31.7	36.2	40.1	44.4	47.8	51.7	61.3	70.4	76.5	82.2	90.1	97.6	104.8	104.9
U.K.	32.7	36.1	39.0	42.0	44.4	47.0	51.4	54.7	57.5	60.3	64.1	68.8	71.8	76.2	81.8	87.9	91.9	96.0
France	24.6	29.8	34.7	36.2	38.4	41.4	45.9	47.9	49.9	50.9	51.5	55.6	61.9	68.5	75.0	80.3	85.9	92.8
Italy	12.4	14.3	15.2	16.8	17.8	19.4	20.8	24.2	26.0	27.7	30.0	33.2	38.0	43.0	47.0	50.7	54.8	59.3
Canada	14.7	17.8	21.4	22.3	22.5	24.3	27.2	29.1	29.8	31.9	31.9	32.3	32.9	35.1	38.4	41.6	46.2	48.2
Netherl.	4.4	5.0	5.3	5.7	6.4	7.2	7.7	8.5	8.8	9.3	10.4	11.5	12.3	13.4	15.7	17.4	18.7	20.6
Belgium	6.9	7.8	7.9	7.5	7.8	8.2	8.8	9.4	9.5	9.7	10.3	10.8	11.6	12.4	14.0	15.3	16.0	17.0
Denmark	2.8	3.0	3.2	3.5	3.6	3.7	4.0	4.2	4.4	4.9	5.2	5.9	6.6	6.9	7.9	8.8	9.6	9.8
Turkey	3.3	4.0	4.6	5.4	5.5	6.8	7.9	7.4	5.5	4.5	4.9	5.0	5.6	6.4	6.9	7.4	8.6	9.5
Norway	2.0	2.4	2.6	2.6	2.8	3.0	3.4	3.7	3.6	3.8	4.0	4.4	4.8	5.1	5.6	6.3	6.8	7.4
Greece	1.7	2.2	2.3	1.9	1.8	2.0	2.3	2.5	2.8	2.9	3.2	3.6	3.8	4.2	4.6	5.2	5.8	6.1
Portugal	1.3	1.4	1.4	1.5	1.6	1.6	1.8	2.0	2.0	2.1	2.3	2.5	2.6	2.8	3.1	3.4	3.7	4.2
Luxemb.	.2	.3	.3	.3	.3	.3	.4	.4	.4	.4	.5	.5	.5	.5	.6	.6	.6	.7
Iceland	.1	.1	.2	.2	.2	.2	.3	.3	.4	.4	.2	.2	.2	.3	.4	.4	.5	.5

Sources for tables 4.1 and 4.2: D/GNP 1950-62 from Ypersele in *Review of Economics and Statistics*, p. 528. Defense expenditures 1963-67 from NATO press release, Jan. 16, 1969. GNP from OECD, *Statistics of National Accounts, 1950-1961* (Paris: OECD, 1964); *National Accounts Statistics, 1955-1964* (Paris: OECD, 1966); *National Accounts of OECD Countries, 1958-1967* (Paris: OECD, 1969). Belgium GNP 1950-52 and Luxembourg 1950-51 calculated from Ypersele D/GNP ratio and defense figures from NATO press release, Dec. 16, 1964. Luxembourg 1966 and 1967 from NATO information service.

Table 4.2. Defense as a Percentage of GNP for all NATO Countries, 1950-67

	1950	51	52	53	54	55	56	57	58	59	60	61	62	63	64	65	66	67
U.S.	5.4	10.9	14.9	14.8	12.9	11.1	10.7	10.9	11.0	10.3	9.9	10.1	10.3	9.6	8.7	8.1	9.1	10.2
W. Ger.*	5.2	6.0	6.5	5.0	4.7	4.8	4.2	4.8	3.4	5.1	4.7	4.7	5.6	6.1	5.4	5.1	4.9	5.1
U.K.	7.3	8.9	11.2	11.3	9.9	9.4	8.8	8.1	7.8	7.4	7.3	7.0	7.2	6.9	6.8	6.7	6.5	6.7
France	6.5	8.4	10.3	11.0	8.7	7.6	9.2	8.7	8.1	7.9	7.6	7.5	7.3	6.8	6.6	6.4	6.3	6.3
Italy	4.6	5.1	5.5	4.6	4.9	4.5	4.5	4.3	4.3	4.2	4.0	3.9	4.0	3.8	3.8	3.8	3.9	3.7
Canada	3.1	6.5	8.9	9.0	8.1	7.6	7.1	6.5	6.0	5.4	5.2	5.3	5.1	4.5	4.4	3.7	3.6	3.7
Netherl.	5.4	5.5	6.2	6.2	6.6	6.2	6.3	5.7	5.0	4.3	4.4	4.9	5.0	4.8	4.7	4.3	4.1	4.3
Belgium	2.4	3.4	5.0	5.2	5.0	4.1	3.8	3.9	3.9	3.7	3.6	3.7	3.7	3.6	3.6	3.3	3.3	3.1
Denmark	1.8	2.3	3.0	3.7	3.6	3.6	3.4	3.5	3.3	2.9	3.1	2.9	3.4	3.5	3.2	3.2	3.1	3.0
Turkey**	6.4	5.8	5.6	5.4	6.0	5.6	5.2	4.5	4.5	5.3	5.5	6.1	5.9	5.5	5.6	5.8	5.2	5.4
Norway	2.6	3.4	4.5	5.7	5.7	4.4	3.9	4.0	4.0	4.1	3.6	3.7	4.1	4.0	3.9	4.2	4.0	4.0
Greece**	7.5	8.1	7.8	6.1	6.5	6.2	7.1	6.0	5.8	5.9	5.9	5.2	4.9	4.3	4.1	4.0	4.1	5.1
Portugal	4.1	3.8	4.1	4.5	4.7	4.7	4.5	4.5	4.5	4.9	4.7	7.2	7.5	7.0	7.2	6.7	6.8	7.9
Luxemb.	1.5	1.7	2.6	3.2	3.6	3.6	2.1	2.2	2.1	1.9	1.1	1.2	1.5	1.4	1.6	1.5	1.5	1.2
Iceland	0.0	0.0	0.0	0.0	0.0	0.0	0.0	0.0	0.0	0.0	0.0	0.0	0.0	0.0	0.0	0.0	0.0	0.0

*Before joining NATO in 1955 the Federal Republic contributed to the defense budgets of some NATO countries by paying occupation costs; it also bore certain other costs falling within the NATO definition of defense expenditures.

**Greece and Turkey joined NATO only in 1952 but were previously under explicit American guarantee with the Truman Doctrine. They and West Germany are included in all computations for all years.

Table 4.3. Measures of Association between D/GNP and GNP and the Mean and Standard Deviation of D/GNP, for NATO, 1950-67

	1950	51	52	53	54	55	56	57	58	59	60	61	62	63	64	65	66	67
Tau (Rank-order)	.42	.65	.73	.60	.59	.62	.55	.61	.52	.53	.54	.41	.48	.47	.45	.46	.42	.39
Percentage of Variance Explained (r^2 x 100)	35	72	81	74	71	69	67	74	68	71	72	59	62	66	58	55	58	51
Percentage of Variance Explained without U.S.	28	51	65	61	45	46	44	51	40	43	38	23	27	33	27	26	25	18
Mean D/GNP without U.S.	4.2	4.9	5.8	5.8	5.6	5.2	5.0	4.8	4.5	4.5	4.3	4.5	4.7	4.4	4.4	4.2	4.1	4.3
Standard Deviation without U.S.	2.3	2.6	3.0	3.0	2.5	2.3	2.5	2.2	2.1	2.0	2.1	2.2	2.1	2.0	2.0	1.9	1.9	2.1

Note: All percentages of variance ⩾21 are statistically significant at the .05 level and ⩾36 at the .01 level, with a one-tailed test. Similarly all taus are significant at the .02 level and ⩾44 at the .01 level.

ances of NATO, including their new alliance with a number of medium powers which, with the United States, could also help carry the burden of Greek and Turkish security. The high measures of association thus achieved in the early 1950s reflect NATO's "shaking down" as it met its primary function of providing the collective good, deterrence to the European states.

When the United States, with the (after 1950) largest D/GNP share, is omitted, the pattern is still much the same, but the degree of relationship is smaller and finally, in 1967, not statistically significant. Even in 1967 the three powers with the largest shares below the United States were Portugal, Britain, and France. The last two were numbers three and four on the GNP rankings, below West Germany, and hence just where the theory would expect them to be—taking up some of the disproportionate burden of the alliance that the United States bore in greater measure. West Germany has always been, for its size, low on the D/GNP scale and near the NATO mean (without the United States). This might be explained by the allies' distaste for too large a German army, the absence of German commitments (providing private goods) elsewhere in the world, and/or by the heavy concentration of NATO ground forces on German soil, providing a special measure of joint defense as well as deterrence. Portugal is now very high because of its colonial wars in Africa, clearly not covered by the collective benefits of NATO. Thus the initial prediction of the theory—that D/GNP and GNP rankings would coincide perfectly—is rejected, but the general pattern holds and the exceptions fit within the broader context of the theory. Similarly, the other states now markedly higher in defense spending than in income are Greece and Turkey, geographically exposed for defense and relatively weakly tied to the other allies by the intangible bonds that help give alliance deterrence its public good quality. Canada, low on exposure and high on intangibles with the United States, is much farther down on the D/GNP list than "pure" public goods theory would have it. Thus the theory both gives us a general perspective on NATO burden-sharing and, by providing an unambiguous "prediction" against which to measure its "failures," allows us better to understand the deviant cases.

If by this test NATO has clearly been a success, it has achieved additional success in providing security for the *non*allied states of Western Europe. Ypersele[16] found, and we can confirm it with slightly later data,[17] that these states rank about where one would expect from the theory. In D/GNP the ordering from high to low is Yugoslavia, Sweden, Spain, Switzerland, Finland, Austria, Ireland; for GNP, Sweden, Spain, Switzerland, Yugoslavia, Austria, Finland, Ireland.[18] Sweden's GNP is about the same as that of NATO-member Netherlands, and its defense ratio is only slightly higher. The worst fit is for Yugoslavia, which is politically and geographically particularly exposed. Thus a formal political bond has not been necessary in order to extend, to a substantial degree, the Western deterrent to cover these nations—a fact of great political importance.[19]

The declining relation over time between GNP and defense shares in NATO also demands careful attention. It largely began with a sharp drop between 1960 and 1961, then attributable to Portugal's jump to third place from a tie with Germany for seventh. But even without Portugal's overseas difficulties, there would have been a modest and fairly steady decline from the peak of the early 1950s. Two at least partially conflicting assessments have been made of contemporary NATO. One stresses the alleged "decline" in the credibility of America's deterrent umbrella for other countries as a result of the achievement of near parity and a secure second-strike force by the Soviet Union. By this view, the European allies have become less confident of the public good provided by NATO and must increasingly fend for themselves militarily. One observer puts the dilemma this way: "The debate on nuclear matters within NATO turns on the issue of confidence. The U.S. tends to ask those of its allies possessing nuclear arsenals: If you trust us, why do you need nuclear weapons of your own? The allies reply: If you trust us, why are you so concerned about our possession of nuclear weapons?"[20] The other view is of course that the Europeans are now either so satisfied with the American guarantee or confident of the reality of general relaxation of East-West tensions that they no longer feel the need to make significant military efforts.[21] Neither is inconsistent with a declining association between defense shares and GNP. On the contrary, for both one would predict that their levels of military effort would increasingly be determined by the private goods defense spending may provide and that the relative importance of the private goods would vary with political, economic, and strategic factors.

The declining credibility thesis, however, would only be consistent with *rising* military budgets in Europe and a narrowing of the gap between the share of income that the superpower United States spends for arms and the share borne by the average NATO ally. (As the United States provided less collective deterrence per dollar its allies would have to spend more.) But this seems not to have happened. With a few minor fluctuations, the average D/GNP ratio for the European allies and Canada remained quite steady after the mid-1950s, then fell noticeably in 1965 and 1966. Once the Korean War effort was over the American military spending ratio also remained quite constant over the same years, with it too beginning to dip in 1963 and later. But from immersion in the Vietnam imbroglio the American military effort rose while almost everyone else's was continuing to decline. After having been as low as 3.9 percent in 1965, the gap between the 1967 United States D/GNP ratio (10.2) and the average for its allies (4.3) was up to 5.9 percent. The gap had been decreasing pretty steadily from 1959 onward (from 6.5 in 1958), but the trend was reversed in 1966. The wider gap may be largely attributable to the war, not shared by Europe, but certainly the fallen D/GNP shares for the allies are not consistent with decreasing confidence in America's deterrent—they are not fending for themselves militarily. France, incidentally, is still only fourth from the top in defense spending—precisely where the theory would place it.

America's NATO allies thus do seem still to be satisfied with the American guarantee, insofar as they perceive a serious external threat at all. Perhaps fallen D/GNP ratios and the declining power of the public goods theory to predict relative contributions among them, however, indicate a greater sense of the futility of collective effort for defense by any European-based (including American) conventional forces. Defense in a ground war, even if "successful," would mean utter devastation for Europe. The efforts of the middle powers—Britain, France, and Germany—may seem decreasingly relevant to the small states. All nations (except Iceland) continue to make some military expenditures for their apparent "private" benefits, and perhaps because they are persuaded that some contribution is required to keep the American five-division "tripwire" on European soil. But as the average D/GNP figure for the other NATO allies has declined, so too has the standard deviation, a measure of the variation of defense spending among them at any point in time. With that index having fallen from 2.5 in 1956 to just 2.1 in 1967 (1.9 in 1966), these states now share similar perceptions of the level of external threat and/or the amount of exertion it is worthwhile to make in order to meet it.[22] In the sense of common perceptions, the allies (excluding the United States) have grown "closer."

Is NATO a "privileged" group, where the largest member would in fact meet the entire cost of deterrence for the alliance even without any contribution from the small states? Possibly so early in the cold war, though the United States was never put to the test by its allies. Now, however, it is doubtful that Congress, at least, is sufficiently persuaded of the need or value of protecting Europe to provide the whole of the collective good. With Europe so recovered economically, perceptions of "fairness" could hardly allow Congress to accept such a burden. Doubtless there would be no formal renunciation of American obligations under NATO but merely a withdrawal of most of the United States troops in Western Europe. It is these troops, unavoidably in the way of any Soviet advance, that so surely make the American deterrent a public good for Europe. In America, as well as in Europe, many now regard a Soviet ground threat to France or Germany as very slight. Europeans who consider the absence of threat to be contingent on the American presence must beware of activating a more radical relaxation in Washington.

We can complete the examination of NATO military spending by looking in another way at the degree to which individual NATO allies agree with the United States or with each other on fluctuations in the level of external threat. Each nation's spending can be correlated with the American D/GNP ratio and with the NATO mean (excluding the United States) over time. That is what is done in table 4.4. One would expect high agreement among "close" allies. Iceland is omitted because its spending is always zero.

Table 4.4 Percentage of Variance in Various NATO Countries' D/GNP Ratios Accounted for by American Spending and by the NATO Average (excluding United States) over the Years 1950-67

	U.S.	NATO Mean
Belgium	88	79
Canada	81	92
France	73	78
U.K.	69	94
Norway	60	52
Luxembourg	43	68
Netherlands	35	74
Italy	32	60
Denmark	29	14
Greece	12	31
West Germany	01	03
Turkey	*07**	00
Portugal	*11**	*27**
U.S.		78

*Negative correlations.

Note: All percentages ⩾17 are statistically significant at the .05 level and all ⩾30 at the .01 level.

There is really a great deal of variation from one nation to another, though the majority of correlations, both with the United States and with the non-American NATO average, are statistically significant (20 out of 27 reach the .05 level). By and large the correlations with

the United States and with the NATO average are about
the same, although the latter are a little higher, reflect-
ing the divergence of the United States especially since
Vietnam. France is much less out of step than is commonly thought. Greece and Turkey are again odd men,
as they were unexpectedly high on the D/GNP listing,
and Portugal, whose African conflicts have made it be-
have so differently. Probably most surprising are the low
correlations for West Germany, so often pictured as
closely tied to the United States and heavily dependent
on NATO protection. As we remarked earlier, the Federal Republic's defense share has changed little in response to any of the cold war's ups and downs.

The Warsaw Treaty Organization

Long-term data on defense expenditures and GNP for
the Warsaw Pact nations are harder to obtain and of
more dubious quality than the figures that are readily
available for NATO countries. The most reliable data are
probably those put together with great care by an East
European specialist, Frederic Pryor, for both variables in
1956 and 1962. More recent figures are those of Taylor
and Hudson for 1965 and the Institute for Strategic
Studies for 1967.[23] The latter two are not fully comparable with the Pryor data but are adequate to give a
general view of the situation. All are shown in table 4.5
and the measures of association in table 4.6.

In the early years of the Warsaw Pact (signed in 1955,
although a fairly complete network of bilateral treaties
existed earlier) the prediction of collective goods theory
strikingly fails of fulfillment. In fact, except for the
leading position of the USSR on both lists, the relationship is negative in 1956 and utterly absent in 1962.[24]

Table 4.5. GNP in Billion $U.S. and D/GNP Percentages for Warsaw Pact Nations, Selected Years 1956-67

	1956		1962		1965		1967	
	GNP	D/GNP	GNP	D/GNP	GNP	D/GNP	GNP	D/GNP
USSR	174.5	12.3	254.8	9.4	313.0	9.0	358.9	9.6
Poland	20.4	4.7	27.9	3.9	30.8	5.5	30.8	5.4
East Germany	20.4	6.0	26.4	3.9	21.5	3.8	28.7	3.7
Czechoslovakia	15.2	5.6	21.4	4.7	22.1	5.9	25.5	5.7
Romania	8.9	5.5	14.6	2.9	14.8	3.4	17.1	3.1
Hungary	8.0*	6.6*	10.8	2.5	11.1	2.7	12.0	2.6
Bulgaria	3.7	9.0	6.1	5.8	6.8	2.9	7.5	3.0

*1955 figures.

Table 4.6. Measures of Association between D/GNP and GNP, and Means and Standard Deviations of D/GNP, for the Warsaw Pact, Selected Years 1956-67

	1956	1962	1965	1967
Tau (Rank-order)	−.14	.29	.81	.81
Percentage of Variance Explained (r^2 x 100)	27	53	89	89
Percentage of Variance Explained without USSR	57*	02*	71	77
Mean D/GNP without USSR	6.2	4.0	4.0	4.0
Standard Deviation without USSR	1.4	1.1	1.2	1.2

*Negative correlations.

Note: All taus >.76 are significant at the .01 level; all percentages for 7 countries >60 are significant at .05 and all >80 at .01; those >66 for 6 countries are significant at the .05 level and >85 at the .01 level.

Nevertheless, by the middle 1960s the predicted positive relationship between size and defense shares appeared, with Warsaw actually outdoing NATO in this respect.[25]

First, let us look at the deviations from the prediction. Bulgaria's high rating is very surprising for such a small nation, but it was geographically exposed—a common border, active border disputes with Yugoslavia and Greece, and no common boundary with the Soviet Union. Among the other countries there is really quite little variation for 1956, except perhaps for the low score of Poland, which was tied with East Germany for second in the economic rankings. Later, as the expected direction of disproportionality begins to appear, the deviant cases are Hungary, consistently at the bottom of the D/GNP list, and East Germany, low for its size. A half-century of war explains Soviet policy toward East Germany, which hosts the largest concentration of Soviet military personnel of any East European state and is not encouraged to build too large an army of its own. The need to appease Hungarian consumers, a legacy from 1956, accounts for that country's laggardness—perhaps along with the dubious reliability of Hungarian troops.

Secondly, we must better understand the failure of collective goods theory to predict the ordering, below the Russians' first rank, in the earlier years. One might of course see it as an extreme case of the recent NATO experience, with so much of the collective good, deterrence, provided by the superpower that all the small and medium-sized Warsaw allies felt secure, and their military expenditures were determined largely by quite different private good considerations. But this fits with neither acknowledged Soviet strategic weakness in the 1950s nor the image of the Warsaw Pact of the 1950s as a closely bound, hierarchically organized, association under the firm control of the Soviet Union. And internal evidence in the data argue strongly against this interpretation too: (1) The mean D/GNP ratio for 1956 even without the Soviet Union is very high, higher than ever reached by the NATO countries and more than a third again as high as NATO in 1956; (2) The standard deviation is appreciably lower than the equivalent for NATO in both 1956 and 1962;[26] (3) Most important, in 1956 the correlation without the Soviet Union is negative, not just absent. It therefore seems likely that coercion was applied by the Soviet Union to extract military contributions from its allies and that this coercion was, as one might suspect, successful in inverse relation to the size of the satellite. Recall that collective goods theory is meant to apply only to voluntary organizations; the Warsaw Pact of a decade ago apparently was voluntary neither in the sense of free choice of membership nor of free choice of contribution level.

We are left then with the evidence that collective goods theory does explain well the pattern of defense expenditures in the middle 1960s. The very high correlation between size and effort among the non-Soviet countries and the widening gap between the Russian D/GNP ratio and the non-Soviet Warsaw mean (up to 5.6 in 1967 from 5.0 in 1965) indicate that insofar as the smaller nations feel a need for military protection, they are now confident in Soviet deterrent strength. To a lesser extent the supplementary security provided by forces of the larger East European nations is important since the ranking is ordered roughly by size and is not random. In at least one sense, then, the Warsaw Treaty Organization has become a voluntary association for Eastern Europe—not voluntary in the sense that members are free to leave but in the sense that they may, in large measure, set their own levels of contribution in accordance with their own perceptions of security needs rather than have their military efforts dictated by the superpower member. The Russians can keep Romania in the alliance but cannot coerce Bucharest into making much of a military effort.[27]

The Soviets, like the Americans, now find themselves with alliance commitments and responsibilities but without the accretion to their own military strength they initially expected. Furthermore, the widened gap between the superpowers' defense ratios and the average for their European allies means that in continuing a policy of armed confrontation both America and Russia have become increasingly isolated from their pact members.

Other Pacts, Various

Four other major multilateral alliances of contemporary international politics explicitly cite security goals for their organizations and include in their charters a statement to the effect that an armed attack against one will be resisted by all.[28] They are SEATO, CENTO (Central Treaty Organization, called Baghdad Pact before the withdrawal of Iraq), the Rio Pact (Inter-American Treaty of Reciprocal Assistance), and the Arab League. The appropriate data and usual measures for 1965 are given in tables 4.7 and 4.8.

SEATO may be viewed in the same light as NATO and contemporary Warsaw, with deterrence provided by the superpower member and, to a much lesser degree, by the middle-sized states of Britain and France. The latter are probably not very important in this context and owe their "correct" rankings more to NATO, although in the mid-1960s the United Kingdom did maintain a significant military presence in Southeast Asia. Primarily it was to protect Malaya and Singapore, but in principle it was available to SEATO members as well. The most notable exception to the predicted rankings would probably be Pakistan, but with a specific conflict situation involving India (noncommunist and therefore excluded from SEATO's concerns) Pakistan's private good needs are readily apparent.

Table 4.7. GNP in Billion $U.S. and D/GNP Percentages for SEATO,
CENTO, Rio Pact, and Arab League Nations, 1965

	GNP	D/GNP
	SEATO	
U.S.	695.6	7.6
U.K.	99.3	5.9
France	94.1	5.5
Australia	22.7	3.7
Pakistan	11.2	5.3
New Zealand	5.23	2.2
Philippines	5.17	1.5
Thailand	3.93	2.2
	CENTO	
U.K.	99.3	5.9
Pakistan	11.2	5.3
Turkey	8.8	5.2
Iran	5.9	5.0
	Rio Pact	
U.S.	695.6	7.6
Brazil	30.0	2.9
Mexico	19.4	.8
Argentina	17.2	1.7
Venezuela	7.69	2.3
Colombia	5.10	2.0
Chile	4.84	2.4
Peru	4.27	2.9
Uruguay	1.56	1.5
Guatemala	1.41	1.0
Ecuador	1.10	2.1
Dominican Republic	.96	3.6
El Salvador	.79	1.3
Panama	.63	.2
Bolivia	.61	2.2
Costa Rica	.59	.3
Nicaragua	.57	1.4
Honduras	.50	1.4
Paraguay	.44	2.9
Haiti	.33	2.8

Table 4.7 (cont.)

	GNP	D/GNP
	Arab League	
Egypt	4.70	8.3
Algeria	2.63	3.8
Morocco	2.61	4.0
Iraq	1.91	10.3
Kuwait	1.58	2.0
Saudi Arabia	1.52	8.6
Sudan	1.36	2.4
Syria	1.13	8.4
Lebanon	1.12	2.7
Tunisia	.95	1.5
Libya	.88	3.2
Jordan	.51	13.0
Yemen	.49	2.2

Source: Taylor and Hudson, *World Handbook.* NATO members' figures differ from NATO and OECD materials used earlier because of different definitions.

CENTO fits the public good prediction perfectly. Whether this is because of security provided by the members themselves or by the latent presence of the United States is unclear from the data, though the

Table 4.8. Measures of Association between D/GNP and GNP, and Means and Standard Deviations of D/GNP, for SEATO, CENTO, Rio Pact, and Arab League, 1965

	SEATO	CENTO	Rio Pact	Arab League
Tau (Rank-order)	.84*	1.00*	.14	.15
Percentage of Variance Explained	83*	99*	38*	02
Percentage of Variance Explained without Superpower	61		00	
Mean D/GNP**	3.8	5.4	1.9	5.4
Standard Deviation**	1.7	0.3	0.9	3.6

*Significant at the .01 level.

**Without U.S. in the case of SEATO and Rio.

American guarantee to these countries must surely be the most important. The United States is an associate member of the pact, is formally allied by other treaties with Britain, Turkey, and Pakistan, and signed a Mutual Defense Assistance treaty with Iran as early as 1950. Mutual Defense Assistance pacts do not carry the explicit guarantees of American alliances, but cover the terms of military aid in arms and advisors. And aid has been formidable: $677 million in military assistance from 1946 to 1964.[29] On the other hand, note the *very* slight variation in D/GNP shares among the four members, as befits an alliance that lacks a superpower and where all members feel threatened. (Pakistan, Turkey, and Iran all have common borders with the Soviet Union.) Also, the mean defense ratio of the CENTO countries is very high—of the four alliances, only for it and the average for the Arab League is the mean very markedly above the average for all the world's nations (3.7 percent).

The Rio Pact situation is much more complex. The United States military expenditure ratio is very much higher than that of any other nation, as would be expected in an alliance where one power is so predominant—with a GNP more than 20 times as large as second-ranking Brazil. For this reason, and for this reason alone, the product-moment correlation for all 20 countries is moderately high and statistically significant.[30] But below the United States, size bears absolutely no relationship to military effort. Furthermore, the average D/GNP ratio for the Latin American countries is very low, a mere 1.9, by far the lowest of any major alliance and much below the world average.[31] One should not conclude from the statistical failure that the alliance collective good of security is not being provided.

Rather, the only conclusion can be that insofar as the Latins perceive a need for protection from extra-hemispheric threats at all, it is being provided in full measure by the United States, leaving no slack or requirement for supplement by the rest of the alliance. The Rio Pact is a privileged group where one member, the United States, finds it worthwhile to provide the entire collective benefit by itself. Thus military expenditures of the other Rio nations are purely private goods.

Why then was the alliance even formed, assuming governments' clear perceptions of their interests at the time? Possibly most important, the western hemisphere nations may have felt that a formal, written commitment to collective defense, most especially such a commitment from the United States, was necessary to make Yankee deterrence credible. Without the explicit pact the public good might not have been provided in sufficient quantity to satisfy some Latins.[32] Desire for group unity, cooperation, and other private goods are also relevant. For the United States, it is said that "alliances have appealed to the American ideal of organizing order by means of collective institutions."[33] A major purpose of the Rio Pact has always been collective security (establishing procedures for the peaceful settlement of disputes among members) as well as collective defense against external threats.[34] The United States has often worked through the Organization of American States in order to have its hemispheric behavior legitimized by joint action. Even without a significant military contribution from its Latin allies, the United States has therefore received important benefits from the pact.

Goals other than simply protecting member states from aggression by a common enemy are apparent from close perusal of the formal charter of this and other alliances, where economic, social, and cultural aims are mentioned and may have helped some states decide to join. If further inducements for Latins to enter or remain in the organization were needed, side-payments, principally economic and military aid from the United States, were always available—as ultimately was the threat of various kinds of sanctions against the intransigent.

Since their military expenditures serve no extra-hemispheric security function for Latin Americans, what are the specific circumstances that make them worthwhile? Particularly for some states there is a certain need for security from other Latin American countries; possibly this is especially important for the Dominican Republic and Haiti, island neighbors of Cuba, which respectively rank first and fifth in D/GNP among the nineteen Latin members of the pact. Counterinsurgency warfare and other internal security functions are clearly important. A very moderate, and not statistically significant, negative rank-order correlation between military effort and per capita GNP is worth noting. But the low mean and standard deviation for Latin American military spending indicates that in the absence of external threat there are rather severe limits to the private goods armed forces can provide to any government in the area.

In sum, the Rio Pact both fits the theory of public goods as commonly applied to alliances and shows some restrictions of the theory. As an example of the extreme provision of a public good by the dominant ally, insofar as there is demand for that good at all, it may in some respects presage the future of NATO and ultimately perhaps even of the Warsaw Pact.

The Arab League evidences no production whatever of a collective good, with negligible correlations between

size and defense shares. No reasonable division of the
entire set improves the fit. Simply ignoring the largest
power, as we did with many other alliances, makes no
difference. One might assume that it would be better if
one excluded the North African members (except
Egypt) on the grounds that they are peripheral to the
common purpose of the alliance, confronting Israel. Or
one might limit the subset still further by excluding all
but those (Egypt, Iraq, Syria, Jordan, and perhaps Leba-
non) directly engaged in conflict with Israel. In both
cases the correlations remain essentially zero. No collective good is being provided—as is further testified by the
very high mean and standard deviation for the League.
Those most in conflict with Israel, regardless of size,
have very high defense ratios; several states distant from
the allegedly common enemy have quite low military
shares. This is a low-grade war, at least a semi-"damage
limitation" situation, which sporadically becomes vio-
lent; for example, of the five states with the highest
defense ratios in 1965, four were the states which engaged in significant conflict in the six-day war two years
later. (The fifth, Saudi Arabia, was in 1965 engaged in
another war, against Egyptian forces in Yemen.) As
noted in the opening discussion, actual defense or war
situations severely limit the theory's applicability, since
defense, unlike deterrence, is rarely much of a public
good between nations.

Two other possible explanations for the inapplica-
bility of the collective goods prediction in this case can
be mentioned. One is that the Arab League has nothing
to do with defense or deterrence but is in fact a purely
aggressive organization, subject to very different explanations. But after watching the Arabs lose three con-
secutive wars with Israel, it seems to us that the Middle

East, rather than being populated by villains and victims, is yet another example of mutual exacerbation. The other, consistent with a modified public goods theory, calls attention to the absence of any very great disparity in size between the largest power and the others, unlike Warsaw or the American-led alliances. Since among nations only a portion of deterrence is truly a public good for others, successful provision of military security to allies may require much greater predominance than any single Arab state can muster.

Alliances and the Goals of American Policy

From a theoretical viewpoint, both the power of the theory of collective goods and its limitations are impressive. Occasionally it works quite well and in so doing illuminates the purposes of an alliance. In other instances it predicts less well or not at all. It should not be taken as a universal key to alliance burden-sharing, as some writers have implied. But the theory's failures, as well as its successes, help to show what are the goals of particular alliances and of particular states. Used carefully it has a very wide potential for explicating international problems. Proponents of collective security might benefit from considering how rarely the conditions for true joint action against a rule-violator are likely to occur. The collective goods problem permeates all of international relations because of the low level of organization or coercion present in the international system.

The theory tells us enough about what one can reasonably expect from various kinds of alliances to offer considerable guidance to Americans. In forming or retaining an alliance with other states and providing them with deterrence, a great power such as the United States must expect that its allies will continue to contribute something to their own military security but less than they would spend in the absence of the alliance. (Unless, perhaps, lacking alliance the small nation would find any military effort within its power to be hopelessly inadequate and hence resort to some sort of unarmed neutrality.) Therefore, the total amount of military spending by states in the alliance will be *less* than would occur without the alliance. If the American goals are to bring other nations under its deterrent shield so as to protect them either more effectively or at less cost to them—allowing them, for instance, to use more scarce resources for development—then it should so judge the alliance's success.

If, however, the United States goal is to save money for itself by in part substituting allies for its own military expenditures, the evaluation is more difficult. America must balance the (small) increment to its own security that will be provided by the collective good portion of the new ally's spending against the additional costs the United States will incur in whatever extra military forces it needs to give security to the ally. Probably neither of these will be great, and by this criterion the net gain or loss from the alliance is not likely to be great.

But neither of these criteria allow for the overall economic and political value of the small ally, beyond its immediate military contribution of forces-in-being.[35] If, in the absence of alliance, the small ally could not or would not spend enough to protect itself, then it is only necessary that the net cost of putting the ally under the great power's umbrella not exceed the additional expenditures for American security that would be required if an opposing power or alliance gained control of the small state. From that perspective, the alliance is much more likely to appear as a benefit to the United States. Insofar as a big power chooses wisely, concentrating on states of high intrinsic value that would not or could not defend themselves adequately alone, the big power will be better off for having accepted the commitment. In any event, a great power such as America or Russia must still expect to spend a greater share of its resources for military security than will small powers. That is the price of primacy.

In evaluating NATO, the verbal balance sheet might look like this: NATO was born when Western perceptions (whether accurate or not) of a communist threat were sharply rising. If one accepts the assumption that some NATO countries would have lost their independence without the American guarantee, then America's acquisition of new allies in 1949 almost certainly kept United States military expenditures from rising much faster and farther than in fact they did. But once the alliance was well established and its forces were built up to meet Europe's security needs reasonably well, it was surely erroneous to hope that American spending could be reduced relative to theirs. Only full political union, making everyone's deterrent spending truly a public good for everyone else, could accomplish that.

Given a constant level of threat to any small nonaligned nation, conclusion of an alliance with it is unlikely to permit any relaxation of United States military expenditures. Any efforts to establish new alliances, as for example in Southeast Asia (unlikely as that may seem in the current domestic and international political climate), should take these facts into account. Finally, efforts to reduce American defense spending by shifting burdens to our current allies are very unlikely to succeed. To extract a very much larger contribution from others we would have to cut our own by so much as really to diminish our allies' sense of security.

5. The Opportunity Costs of American Defense

Who Pays for Defense?

From the analysis of chapter 3 we have some idea of
what parts of the country benefit from high levels of
defense spending. Quite a number of sound and well-
documented studies of the past few years have located
many groups in the national economy that particularly
profit from military expenditures.[1] They show very
effectively which industries and which states gain disproportionately from defense spending and hence develop some special interest in maintaining or increasing
such expenditures. One need not accept Marxist or
other economic determinist positions on the causes of
war to find such information relevant to identifying
political pressure groups that must be countered or com-
pensated in any effort to reduce the level of military
spending.

A question closely related to "Who benefits from defense spending?" is, of course, "Who *pays* for it?"; but
curiously this second problem has received very little
attention. Nothing comes free, and defense is no exception. In this chapter we shall examine some evidence
about what segments of the economy and society sacrifice disproportionately when defense spending rises.
What kinds of public and private expenditure are diminished or fail to grow at previously established rates,
when military expenditures take a larger proportion of a
less than infinitely expansible economy? The exercise
will use economic data to address some critical political
questions. It should point to particular interests or pressure groups that are relatively strong (or weak) and able
(or unable) to maintain their accustomed standards of

living during periods of international adversity, or to seize the pecuniary opportunities presented by the reduced defense effort that may accompany a relaxation of global tensions. Furthermore, it will enable us, in a sense, to do a cost-benefit analysis of war or preparedness, to identify the opportunity costs in the kind and amount of social benefits that are likely to be forgone. The costs may be entirely in the form of current benefits forgone or, if the nation's resource base is eroded, they may be paid largely by future generations.

The basic question has two prongs, of which only one can be considered here. The other would require detailed data on tax incidences and on wage and price changes. Expanded defense needs are usually financed by a combination of increased taxation and deficit spending. Tax rates would tell us what income or occupational groups suffered disproportionately from assessment increases. Deficit spending in the absence of adequate tax increases normally produces inflation. The wage and price data would show which groups saw inflationary pressures diminishing their real income more sharply than that of the average consumer. Generally one would expect owners of common stocks and land, and union laborers with cost-of-living clauses in their contracts, to suffer least. White-collar workers and especially pensioners and poor unorganized laborers have the most nearly fixed incomes and the greatest vulnerability.

A full examination of these aspects requires a higher degree of data specificity and precision than is currently available. Rather than give unreliable answers we will not develop that side of the problem in detail, though there are some good data on a couple of intriguing aspects. One is the cost of defense to those it affects most directly—armed forces personnel. In shooting wars some of them pay with their blood, but even many noncombatants pay a heavy economic price. All those who serve in the armed forces because of government coercion, rather than out of free choice of the military for its benefits, pay an implicit tax. This is the case both for draftees and for the reluctant volunteers who enlist only to avoid the draft. According to several recent studies of the situation in the 1960s, that "tax" amounts to at least 50 percent of what their income would be as civilians. That is, in the aggregate they give up, during their two to four years in the military, about half their normal income.[2] Similarly, the civilian economy is deprived of services for which it would be prepared to pay that much. Overall, the cost of the draft has been estimated at somewhere between five and ten billion dollars per year.[3] Thus the cost of defense to the country as a whole is really higher than it appears from the Defense Department budget. Our taxes are five to ten billion dollars lower than they would be with the high military pay scale necessary to attract a volunteer army of equal size, but the loss of productivity to the economy is still present.

The regional impact of defense spending deserves just a bit more attention than it received in chapter 3. Not only does defense procurement benefit some regions more than others, it has a regressive effect on the nation's income structure. According to Assistant Secretary of the Treasury Murray Weidenbaum's review of the data, the relatively high-income regions—the Far West and the Northeast—usually receive shares of defense work that are above their shares of the national income. Similarly, the medium- or average-income regions—the Great Lakes, Plains, and Rocky Mountain states—receive shares of defense work that are below their national income shares. For the Southeast and Southwest there is some ambiguity. According to some estimates these lowest-income states also receive less than their income share of defense work; according to others (including the data of chapter 3 here) many of them now get slightly more, which indicates a mild tendency toward the reduction of income inequality at the lower extremity. Overall, however, there is an income redistribution effect in favor of the rich states. That effect is stronger with NASA spending, where all regions except the Far West and Plains states have received a less than proportionate share. Furthermore, Weidenbaum shows that of all major federal expenditure programs, only DoD and NASA spending had this regressive effect —but that the regressiveness was strong enough to wipe out the progressive income redistribution of all other federal programs.[4] Other studies show that even within industry and employment groups, defense spending makes the rich richer. The reason is the need in most weapons manufacture for a highly skilled and therefore expensive labor force.[5]

But instead of pursuing these questions further, in the rest of this chapter we shall examine information on expenditures by GNP categories, by function, and by governmental unit to see what kinds of alternative spending bear the brunt of heavy military spending. For the United States we have this data for the period 1939-68 or 1938-67. This allows us to see the effects of two earlier wars (World War II and the Korean War) as well as the burdens of the current Vietnam venture. In the following chapter we shall compare the American experience with that of several other developed Western states.

First, an overview of the changing level of defense expenditures may be helpful. For 1939, in what was in many ways the last peacetime year this nation has experienced, defense expenditures were under $1.3 billion. They rose rapidly with the new preparedness to a still unsurpassed peak of $87.4 billion in 1944. The 1968 figure (annual estimate based on the first three quarters) was by contract $78.4 billion, reflecting a buildup, for the Vietnam war, from levels of around $50 billion in the early-to-mid-1960s. The raw dollar figures, however, are deceptive because they reflect neither inflation nor the steady growth in the economy's productive capacity that makes a constant defense budget, even in price-adjusted dollars, a diminishing burden. Figure 5.1 shows the trend of military expenditures as a percentage of gross national product over the past thirty years.

We immediately see the great burdens of World War II, followed by a drop to a floor considerably above that of the 1930s. The cold war, and particularly the Korean action, produced another upsurge in the early 1950s to a level that, while substantial, was by no means the equal of that in the Second World War. This too trailed downward after the immediate emergency was past, though again it did not retreat to the previous floor. Not since the beginning of the cold war has the military accounted for notably less than 5 percent of this country's GNP; not since Korea has it had as little as 7 percent. Finally, we see the effect of the Vietnam buildup, moving from a recent low of 7.3 percent in 1965 to 9.2 percent in 1968. This last looks modest enough and is, when compared to the effects of this nation's two previous major wars. At the same time, it also represents a real sacrifice by other portions of the economy. The 1968 GNP of the United States was well in excess of

$800 billion; if we were to assume that the war effort accounted for about 2 percent of that (roughly the difference between 7.3 percent and 9.2 percent) the dollar amount is approximately $16 billion. That is in fact too low a figure, since some billions were already

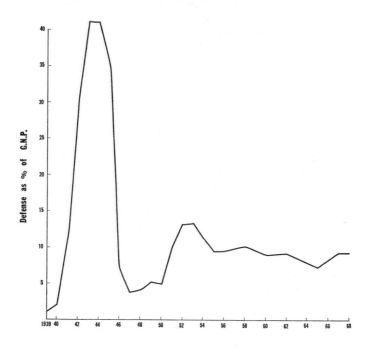

Figure 5.1 Defense Expenditures as a Percentage of U.S. GNP, 1939-68
Source: See table 5.1.

being devoted to the war in 1965, and direct estimates of the war's cost are typically about $25 to $30 billion.[6] The amounts in question, representing scarce resources that might be put to alternative uses, are not trivial.

Criteria for Evaluation

I assume that defense spending has to come at the expense of something else. In the formal sense of proportions of the GNP that is surely true, but it is usually true in a more interesting sense as well. Economics is said to be the study of the allocation of scarce resources; despite some periods of slack at the beginning of wartime periods (1940-41 and 1950), resources have generally been truly scarce during America's wars. Major civilian expenditures have not only lost ground proportionately (as would also happen from a military spending program financed entirely out of slack) but have failed to grow at their accustomed rates, have lost ground in constant dollars as a result of inflation, or have even declined absolutely in current dollars. During World War II, for example, such major categories as personal consumption of durable goods, fixed investment, federal purchases of nonmilitary goods and services, and state and local expenditures all declined sharply in absolute dollar amounts despite an inflation of nearly 8 percent a year. As indicated in chapter 1, there is little evidence that military spending is required to maintain overall demand in the modern American economy. On the contrary, because of the skills with which economic systems are now managed, in modern economies defense expenditures are much more likely to force tradeoffs than they were some thirty years ago. Thus the formulation, "Who pays for defense?" is not inappropriate.

Defense expenditures are not necessarily without broader social utility. The education, skills, and conditioning that young men, perhaps especially minority groups, receive during service in the armed forces are

likely to benefit them and their society when they return to civilian life. Spending for military research and
development produces technological spillovers into the
civilian sector. Rarely, however, is the achievement of
such benefits as spillovers the most efficient way to
obtain them. In addition to learning electronics, soldiers learn combat skills—of dubious benefit to civil
peace when they become civilians again. While scientific
research may be serendipitous, the odds are better that a
new treatment for cancer will come from medical research than from work on missile systems.

Research spillovers are especially likely to be exaggerated. While one can point to some, impressions are
that the important ones are not numerous. Ralph
Lapp notes that when the aircraft companies reached
the development stage for the supersonic transport
they found that, contrary to everyone's expectations,
they could only rarely borrow military technology as
money-saving shortcuts.[7] DoD contractors themselves—perhaps because of their low-unit, high reliability, high-cost production methods—are rarely able to
achieve commercial benefit from their developments.
Weidenbaum found that

> Other than the few firms selling equipment to the
> airlines, the large defense suppliers obtain only 1 or
> 2 percent of their sales from products based on their
> defense/space work that are sold in commercial mar-
> kets. The list of abandoned commercial ventures is
> long, ranging from stainless-steel caskets to powered
> wheelbarrows to garbage-reduction machinery.[8]

The scarcity of important commercial applications
becomes astonishing when one recalls the magnitude
of defense research and development. According to

Weidenbaum's testimony to Congress, in 1963 approximately 52 percent of all engineers and scientists were working on projects financed by defense or space programs.[9] One observer declares that "It would be absurd to believe that a majority or even a substantial minority of the nation's technological talent could long be devoted to weaponry without any sacrifice of progress in civilian sectors of the economy."[10] Though there doubtless are civilian benefits, they can hardly be equal to the resources spent for defense and space. Hence, when in the following analysis we find tradeoffs between defense and spending in other functional categories, we shall still consider the losses as real costs, if not quite as heavy costs as a literal interpretation of the dollar amounts would imply. Defense R & D is not worthless to civilian society, it just is not nearly as productive as directly civil R & D would be—assuming, of course, that support for civil R & D would be forthcoming.

This last point reminds us that some civilian expenditures, for health, for education, and for research, have been stiumlated by cold war and ultimately military requirements. Various programs of the 1950s fit this characterization, when a need more for long-run girding of the loins than for immediate military capabilities was widely seen. Still, this is appropriate to precisely the kind of question we shall be asking. If the civilian and military expenditures consistently compete for scarce resources, they will be highly negatively correlated; if they both are driven by the same demands, they will be positively correlated. If they generally compete but sometimes are viewed as complementary, the negative correlation will be fairly low.

An evaluation of the experience of this and other

nations requires some explicit criteria. There is room for serious argument about what these criteria should be, but I will suggest the following:

1. It is bad to sacrifice future productivity and resources for current defense or war-fighting activities; insofar as possible such activities "should" be financed out of current consumption. Such an assumption might be easily challenged if it were offered as a universal, but for the developed countries of North America and Western Europe in recent years it seems defensible. All of them are now, relative to their own past and to other nations' present, extremely affluent with a high proportion of their resources flowing into consumption in the private sector. Furthermore, for the years being analyzed the demands of defense have not usually been terribly great. Since World War II ended, none of these countries has had to devote more than about 10 percent of its GNP to military needs, save for the United States during the Korean War when the figure rose to just over 13 percent. It surely is arguable that such needs rarely require substantial mortgaging of a nation's future.

a. By this criterion one would hope to see periodic upswings in defense requirements financed largely out of personal consumption, with capital formation and such social investment in the public sector as health and education being insensitive to military demands.

b. Another aspect of our criterion, however, is the anticipation that in periods of declining military needs the released resources would be kept largely for investment and education rather than returned to private consumption. In a strong form the criterion calls for a long-term secular increase in the proportion of GNP devoted to various forms of material and social investment. This secular trend would be realized through a fluctuating line made up of a series of upward slopes followed by plateaus, insensitive to rising defense needs but responsive to the opportunities provided by relaxations in the armament pace. It implies a public-sector approach to this country's newly manifest domestic needs.

2. Another point of view, partially in conflict with the last comment, would stress the need for a high degree of insulation from political shocks. A constant and enlarging commitment to the system's social resources is necessary for the most orderly and efficient growth of the system, avoiding the digestive problems produced by alternate feast and famine. Some spending, for instance capital expenditures for buildings, may be only temporarily postponed in periods of fiscal stringency and may bounce back to a higher level when the pressure of defense needs is eased. To that degree the damage would be reduced but not eliminated. In the first place, school construction that is "merely" postponed four years will come in time to help some students, but one age group simply loses out. Secondly, boom and bust fluctuations, even if they do average out to the socially desired dollar level, are likely to be inefficient and produce less real output than would a steadier effort.

Guns, Butter, and Structures

The method of analysis is straightforward. I have computed various kinds of private and public expenditures as proportions of GNP and correlated them with the proportion of GNP represented by defense spending. Both linear and curvilinear regressions were tried in case the relationship changed at very high defense levels, but the curvilinear fit will be reported only when it improves notably on the results of the linear model. Defense is the "independent" factor, with the implication that the others are dependent upon it; that is, that increases in civil expenditures are made possible by relative reductions in defense spending and that increases in defense must force relative reductions, whether deliberate or unintended, in nonmilitary items.[11] The causal chain is of course not fully demonstrable with regression analysis but seems generally plausible.

Calculation of a nation's GNP is an exercise in accounting; economists define the gross national product as the sum of expenditures for personal consumption, investment or capital formation, government purchases of goods and services, and net foreign trade (exports minus imports). Each of these categories can be broken down further. Private consumption is summed from expenditures on durable goods (e.g. automobiles, furniture, appliances), nondurables (e.g. food, clothing, fuel), and services (airline tickets, haircuts, entertainment); investment includes fixed investment in nonresidential structures, producers' durable equipment (e.g. machinery), residential structures, and the accumulation or drawing down of stocks (inventories); government purchases include both civil and military expenditures of the federal government and spending by state and local units of government. Except for inventories (which fluctuate widely in response to current conditions and are of little interest for this study), we shall look at all these and later at a further breakdown of public expenditures by level and function.

Table 1 gives the percentage of variance in each item of civil spending that is accounted for by defense, the regression coefficients, and an index of proportionate reduction. The percentage of variance tells *how closely* the two expenditure categories vary together, and the regression coefficient tells *the amount in dollars* by which an item of civilian spending changes in response to a one-dollar increase in defense. The proportionate reduction index shows the damage suffered by each category relative to its "normal" base. It assumes for illustration a total GNP of $400 billion and an increase of $25 billion in defense spending from the previous period, and it assumes that the "dependent" expenditure category had previously been at that level represented by its mean percentage of GNP over the 1939-68 period. This last measure is important for policy purposes, since the *impact* of the same dollar reduction will be far greater to a $100 billion investment program than to a $500 billion total for consumer spending.[12] I will also report the examination of year-to-year changes that may be disguised by the overall summary statistics. For instance, some kinds of politically well-entrenched spending may hold their own when defense rises but go up rapidly when military demands relax. The correlation will in such cases be fairly low, though for investment this would in fact meet criterion number 1 above.

There is a widespread myth in America that defense spending, even if higher than strictly necessary for national security, is not really a waste because the alternative use for resources would be equally frivolous civilian spending. That is, if the money were not going for weapons it would be turned back to the taxpayers for personal consumption, not used in the public sector for education, pollution control, urban development, or public health programs. If that were true, then the real price of defense would be only the loss of some luxuries for middle- and upper-class America, and we might well pre-

fer always to err on the side of too much defense spending rather than too little. Certainly congressional behavior with the 1969 tax reform and tax reduction bill did
little to ease the impression that among legislators tax
cuts had higher priority than social needs.

*Table 5.1. The Effect of Defense Spending on Civilian
Activities in the United States, 1939-68*

	Percentage of Variance Explained (linear reg.)	Regression Coefficient	Index of Proportionate Reduction
Personal Consumption (Total)	*84**	*−.420*	*−.041*
Durable Goods	78*	−.163	−.123
Nondurable Goods	04	−.071	−.014
Services	55*	−.187	−.050
Fixed Investment (Total)	*72**	*−.292*	*−.144*
Nonresidential Structures	62*	−.068	−.140
Producers' Durable Equipment	71*	−.110	−.123
Residential Structures	60*	−.114	−.176
Exports	67*	−.097	−.115
Imports	19	−.025	−.037
Federal Civil Purchases	38*	−.048	−.159
State & Local Gov't Consumption	38*	−.128	−.105

*Indicates both the percentage of variance and the regression coefficient are in
principle statistically significant at the .001 level with a one-tailed test. In fact,
because of various technical problems (autocorrelation and non-normal distributions) the significance level is somewhat exaggerated here, but it is nevertheless
high enough to indicate relationships of real interest. See Russett, "Some Decisions in the Regression Analysis of Time-Series Data." in Bernd, ed. *Mathematical
Applications in Political Science, V.*

Source: Survey of Current Business, 45, no. 8 (August 1965: 24-25, 46, no. 11
(November 1966): S-1, and 47, no. 11 (November 1968): S-1. Data for 1968 are
provisional figures for first three quarters only. The defense percentages differ
slightly but not importantly from the standardized NATO-definition figures used
in chapter 4.

As we shall see, however, the truth differs substantially from the myth. The second column of table 5.1,

with the regression coefficients, shows the alternative uses for a defense dollar—from where, in periods of rising defense needs the resources for military goods were extracted or, when military spending was reduced, where in the civilian economy the funds went. Typically it was as follows: 42 cents from personal consumption, 29 cents from fixed capital formation, 10 cents from exports, 5 cents from federal government civilian programs, and 13 cents from state and local governments' activities.

Private consumption has indeed been the largest alternative use of defense money. Guns do come partly at the expense of butter. Changes in defense expenditure account for 84 percent of the variance in total consumption, and the regression coefficient is a relatively high −.420.

Of the subcategories, sales of consumer durables are most vulnerable, with 78 percent of their variance accounted for by defense. Spending on services is also fairly vulnerable to defense expenditures, with the latter accounting for 54 percent of the variance. But the linear regression between defense and nondurables is not nearly so high, with but 4 percent of the variance accounted for. This is not surprising, as needs for nondurables are virtually by definition the least easily postponed. A look at the plot of defense against nondurable consumer goods purchases, however, shows that the proportion of GNP spent on the latter *increased* somewhat in the early 1940s, when war costs were reaching their peak; it was as high in 1942-45 as in many prewar and postwar years. The explanation is fairly simple: during the war years new consumer durables such as automobiles and appliances were virtually unavailable, since the factories that normally produced them were then turning out war ma-

teriel. Similarly, due to manpower shortages almost all
services were expensive and in short supply, and long-
distance travel was particularly discouraged. ("Is this
trip necessary?") Hence, to the degree consumers'
spending power was not mopped up by taxes or saved,
an unusually high proportion was likely to go into nondurables.

Severe restriction of consumption by military demands was forecast by Harold Lasswell in his famous
article on the "garrison state":

> The rulers of the garrison state will . . . most as-
> suredly prevent full utilization of modern productive
> capacity for nonmilitary consumption purchases. The
> elite of the garrison state will have a professional inter-
> est in multiplying gadgets specialized to acts of vio-
> lence. The rulers of the garrison state will depend
> upon war scares as a means of maintaining popular
> willingness to forego immediate consumption . . . If
> ever there is a rise in the production of nonmilitary
> consumption goods, despite the amount of energy di-
> rected toward the production of military equipment,
> the ruling class will feel itself endangered by the grow-
> ing "frivolousness" of the community.[13]

Nevertheless, Lasswell's garrison-state construct is not
generally an accurate representation of contemporary
American society, whatever the latter's faults. A recent
Soviet analysis declares that both consumption and investment will be damaged, perhaps especially the latter:

> In an expanding wartime economy the military
> product also constitutes part of the accumulation
> fund and this serves to diminish the rate and scale or
> reproduction of civilian consumer goods. Thus, mili-

tarization of the economy reduces both personal consumption and real accumulation of capital for the production of civilian goods.

It is possible to manipulate the accumulation fund, i.e., to put an increasing portion of it into economic mobilization, thereby reducing investments in the civil sector of the national economy. This approach is also fraught with the danger that the nation's economic potential will be weakened. Excessive reductions in capital investment have a retarding effect on overall production growth and on the development of productive capacities.[14]

Our data show that investment (fixed capital formation) is typically hard hit by American war efforts; with its consequence of a smaller productive capacity in later years, diminished investment is a particularly costly loss. The 72 percent of variance explained is only a little less than that for defense on consumption, and the regression coefficient is a substantial −.292.[15] The regression coefficient is of course much lower than that for defense and consumption (29 cents on the dollar, compared with 42 cents), but that is very deceptive considering the "normal" base from which each starts. Over the thirty years for which we have data, the mean percentage of GNP going to consumption typically was about five times as great as investment. Thus, in our hypothetical illustration a $25 billion increase in defense costs in a GNP of $400 billion would, ceteris paribus, result in a drop in consumption from approximately $256 billion to roughly $245, or only a little over 4 percent of total consumption. Investment, on the other hand, would typically fall from $51 billion to about $44 billion, or more than 14 percent. For many purposes, therefore, the third column of table 5.1 is the most interesting.

Proportionately then, investment is much harder hit by an expansion of the military establishment than is consumption. Since future production is dependent upon current investment, the economy's future resources and power base are in a very real sense much more severely damaged by the decision to build or employ current military power than is current indulgence. According to some rough estimates, the marginal productivity of capital in the United States is between 20 and 25 percent; that is, an additional dollar of investment in any single year will produce 20-25 cents of annual additional production in perpetuity.[16] Hence if an extra billion dollars of defense in one year reduced investment by $292 million, thenceforth the level of output in the economy would be *permanently* diminished by on the order of $65 million per year. On the trend, investment has fallen from over 15 percent of the GNP in the middle 1950s to under 14 percent.

This position is modified slightly by the detailed breakdown of investment categories. Residential structures (housing) shows the least explained variance of the three, but its regression coefficient is the strongest, and it takes the greatest proportionate damage. Within the general category of investment, therefore, nonresidential structures and equipment usually hold up somewhat better proportionately than does housing. Doubtless this is the result of deliberate public policy, which raises home interest rates and limits the availability of mortgages while trying at the same time to maintain an adequate flow of capital to those firms needing to convert or expand into military production. One other feature should be noticed, however. The curvilinear model (second-order curve) adds as much as 43 percent of the remaining variance with nonresidential structures. Relative spending goes up with increments to very low levels of defense spending to where defense equals about 9 percent of GNP. It then flattens out until, by around the 13 percent mark, it begins to fall off sharply. This phenomenon is probably the result of construction of new factories for war production and, especially, facilities for military bases and camps when a large army must be mustered quickly from low manpower levels.

The nation's international balance of payments is often a major casualty of sharp increases in military expenditures; the present situation is not unusual. Some potential exports are diverted to satisfy internal demand, others are lost because domestic inflation raises costs to the point that the goods are priced out of the world market. Imports may rise directly to meet the armed forces' procurement needs—goods purchased abroad to fill local American military requirements show up as imports to the national economy—and other imports rise indirectly because of domestic demand. Some goods normally purchased from domestic suppliers are not available in sufficient quantities; others, because of inflation, become priced above imported goods. If the present situation is typical, the Vietnam War's cost to the civilian economy would be responsible for a loss of more than $1.5 billion in exports.

The import picture is more complicated. According to the sketch above, imports should rise with defense spending, but the percent of variance explained in the table is very low, and the regression coefficient is actually negative. This, however, is deceptive and gives a conclusion quickly reversed by a curvilinear regression. The curve shows the expectable sharp positive relation between defense and imports up to a level of defense more or less equal to 13 percent of GNP. Only then does it flatten out and then turn down dramatically. The four years of World War II show unusually low importation activity, and the reasons are obvious. A combination of enemy occupation of normal sources of goods for the U.S., enemy surface and submarine activity in the sea lanes, and the diversion of our allies' normal export industries to serve their needs drastically reduced America's opportunity to import. To assess the impact of defense expenditures on imports in a less than global war one must omit the World War II data from the analysis. Doing so produces the expected positive regression coefficient, on the order of +.060. (See figure 5.2.) This suggests that the current effect of Vietnam may be to add, directly and indirectly, over $1 billion to the nation's annual import bill.[17] Coupled with the loss of exports, the total damage to the balance of payments on current account (excluding capital transfers) is in the range $2.5-3 billion. That still does not account for the entire balance of payments deficit that the United States has experienced (twice in recent years it exceeded 3 billion annually), but it goes a long way to explain it.

The Public Sector

In the aggregate there is no very strong impact of defense on civil public expenditures. The variance explained by the linear model is a comparatively low 38; the regression coefficients are only −.048 for federal civil purchases and −.013 for state and local governments. For the federal government a curvilinear regression helps, however. During the four peak years of World War II, changes in federal civil expenditures were

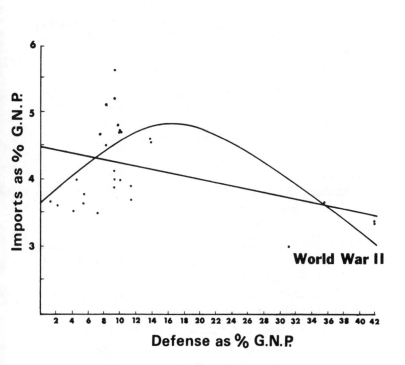

Figure 5.2 Defense Spending and Imports as a Percentage of GNP, 1939-68

essentially unrelated to changes in defense spending, hence the curve at that end is flattened out. But it is simultaneously steepened for the 1 percent to 15 percent defense range.[18] Samuel Huntington notes that "Many programs in agriculture, natural resources, labor and welfare dated back to the 1930's or middle 1940's. By the mid-1950's they had become accepted responsibilities of the government," and hence politically resistant to the arms squeeze. If so, the overall inverse relationship we do find may be masking sharper changes in some of the less well-entrenched subcategories of central government budgeting.[19] We shall investigate this below. Even so, we can see a drop in federal civil purchases from 4.3 percent of GNP in 1939 to about 2.4 percent in the middle and late 1960s. That is hardly consistent with the conservative spectre of big (civilian) government.

Some caution is required when relating state and local government expenditures to defense. There really is no relationship except between the points above and below the 15 percent mark for defense. During World War II state and local government units did have their spending activities curtailed, but overall they have not been noticeably affected by defense purchases. Quite to the contrary, spending by state and local political units for all purposes combined has risen steadily, in an almost unbroken line, since 1944. The rise, from 3.6 percent of the GNP to 11.2 percent, has continued essentially heedless of increases or diminution in the military's demands on the economy.[20]

We must, however, look at the breakdowns by function, since so many observers believe that some government programs are especially vulnerable. Senator Fulbright, for instance, as long ago as 1964 castigated the readiness with which the American people have consented to defer programs for their welfare and happiness in favor of costly military and space programs. Indeed, if the Congress accurately reflects the temper of the country, then the American people are not only willing, they are eager to sacrifice education and urban renewal and public health programs—to say nothing of foreign aid—to the requirements of the armed forces and the space agency.[21]

When we do look at the breakdowns by function it becomes clear that the effect of defense fluctuations is serious, if less distinct than for GNP categories. Three major items—education, health, and welfare—were selected for further analysis, on the grounds that one might reasonably hypothesize for each that expenditure levels would be sensitive to military needs and, for the first two, that a neglect of them would do serious long-term damage to the economy and social system of the nation.

All three are sensitive to defense spending, with welfare not surprisingly somewhat more so than the others. In most of this analysis reductions in expenditure levels that are forced by expanded defense activities represent a cost to the economic and social system, but welfare is different. Insofar as the needs for welfare, rather than simply the resources allocated to it, are reduced, one cannot properly speak of a cost to the economy. Rather, if one's social preferences are for work instead of welfare, the shift represents a gain to the system. Heavy increases in military pay and procurement do mean a reduction in unemployment, and military cutbacks are often associated with at least temporary or local unemployment. The effect seems strongest on

*Table 5.2. The Effect of Defense Spending on Public Civil Activities
in the United States, Fiscal Years 1938-67*

	Percentage of Variance Explained (linear reg.)	Regression Coefficient	Index of Proportionate Reduction
Education (Total)	*35**	*−.077*	*−.139*
Institutions of Higher Ed.	12	−.013	−.146
Local Schools	34*	−.053	−.125
Other Ed.	19	−.014	−.265
Federal Direct to Ed.	16	−.013	−.309
Federal Aid to State & Local Gov'ts for Ed.	08	−.004	−.140
State & Local Gov't for Ed.	24	−.060	−.124
Health & Hospitals (Total)	*32**	*−.017*	*−.113*
Total Hospitals	*30*	*−.014*	*−.123*
Fed. for Hospitals	25	−.004	−.130
State & Local for Hospitals	29	−.011	−.120
Total Other Health	*22*	*−.033*	*−.087*
Fed. for Health	06	−.001	−.101
State & Local for Health	45*	−.002	−.078
Welfare (Total)	*54**	*−.019*	*−.128*
Fed. Direct for Welfare	13	−.003	−.493
Fed. Aid to State & Local Gov'ts for Welfare	17	−.005	−.087
State & Local for Welfare	30	−.011	−.134

*Both percentage of variance and regression coefficient are statistically significant at the .001 level.

Source: GNP same as for table 5.1, but adjusted for fiscal year. Others: U.S. Bureau of the Census, *Historical Statistics of the United States, Colonial Times to 1957* (Washington: U.S. Government Printing Office, 1960), pp. 719, 723-27; *Statistical Abstract of the United States, 1963* (Washington: U.S. Government Printing Office, 1963), pp. 392, 419, 422; *Statistical Abstract of the United States, 1967* (Washington: U.S. Government Printing Office, 1967), pp. 390, 421, 423; U.S. Bureau of the Census, *Historical Statistics of the United States, Colonial Times to 1957; Continuation to 1962 and Revisions* (Washington: U.S. Government Printing Office, 1965), pp. 98-101, 148-50. Budget data for 1967 from U.S. Bureau of the Census, *Governmental Finances in 1966-67,* Series GF67-No. 3 (Washington: U.S. Government Printing Office, 1968), pp. 15, 17-19. Data for Institutions of Higher Education, Local Schools, Other Education, Total, Federal, and State and Local for both Health and Hospitals include 1938, 1940, 1942, 1944, 1946, 1948, 1950, 1952-67; others same except for addition of 1951.

state and local governments' welfare spending. In fact, the inverse relationship between defense and welfare at most spending levels is understated by the linear regression model. When a curve is applied for total welfare expenditures, 32 percent of the remaining variance is added, to make a total of more than two-thirds of the variance. At all but the highest levels of defense spending, achieved in World War II, the inverse relationship is very steep, with small increases in military needs having a very marked dampening effect on welfare costs. But manpower was quite fully employed during all the years of major effort in World War II, so ups and downs in defense needs during 1942-45 had little effect, and the curve flattens out.

Both for education and for health and hospitals their relationship to the immediate requirements of national defense is less powerful (less variance explained) but nonetheless important. Furthermore, the regression line is quite steep for education, and, since the mean share of GNP going to education is only 3.5 percent for the period under consideration, the proportionate impact of reductions is severe.

A widespread assumption holds that public expenditures on education have experienced a long-term secular growth in the United States. That assumption is correct only with modifications. The proportion of GNP devoted to public education has increased by three-quarters over the period, from 3 percent in 1938 to 5.3 percent in 1967. But it has by no means been a smooth and steady upward climb. World War II cut deeply into educational resources, dropping the educational percentage of GNP to 1.4 in 1944; only in 1950 did it recover to a level (3.6) notably above that of the 1930s. Just at that point the Korean War intervened and education once more suffered, not again surpassing the 3.6 level before 1959. Since then, however, it has grown fairly steadily without being adversely affected by the relatively modest rises in defense spending. Actually, educational needs may have benefited somewhat from the overall decline in the military proportion of the economy achieved between the late 1950s and mid-1960s. The sensitivity of educational expenditures to military needs is nevertheless much more marked on the latter's upswings than on its declines. Education suffered immediately when the military needed to expand sharply for World War II and Korea; it recovered its share only slowly after defense spending had peaked. Surprisingly, federal educational expenditures show less variance explained than does spending by state and local units of government; also, local schools at the primary and secondary level are more sensitive than are public institutions of higher education (whose share has grown in every year since 1953).

Public expenditures for health and hospitals are only a little less sensitive to the pressures of defense than are dollars for education. Here again the image of a long-term secular growth deceptively hides an equally significant pattern of swings. They accounted for a total of .77 percent of GNP in 1938; as with education this was sharply cut by World War II and was not substantially surpassed (at 1 percent) until 1950. Once more they lost out to the exigencies of defense in the early 1950s, and bounced back slowly, at the same rate as did education, to recover the 1950 level in 1958. Since then they have continued growing slowly, with a peak of 1.23 in 1967. Thus, the pattern of health and hospitals is almost identical to that for education—some long-term growth but great cutbacks in periods of *heavy* military need and only slow recovery thereafter. In detail by political unit the picture is also much the same—despite reasonable a priori expectation, federal spending for this item is less closely tied to the defense budget than is that by state and local governments. It should also be noted that, though the percentage of variance explained is much the same, the impact of defense on health and hospitals is slightly less severe than on education.

It seems fair to conclude from these data that America's most expensive wars have severely hampered the nation in its attempt to build a healthier and better-educated citizenry. (One analyst estimates that what *was* done to strengthen education accounted for nearly half of the United States per capita income growth between 1929 and 1957.)[22] A long-term effort has been made, and with notable results, but typically it has been badly cut back whenever military needs pressed abnormally hard.

It is too soon to know how damaging the Vietnam War will be, but in view of past regularities one would anticipate significant costs. The inability to make investments will leave the nation poorer, more ignorant, and less healthy than would otherwise be the case. We can already see the effect of the war on fixed capital formation, discussed earlier. Consumption absorbed a larger absolute decline in its share of GNP between 1965 and 1968 than did physical investment—from 63.3 to 62.1 percent in the first instance, from 14.3 to 13.8 percent in the second, but given the much smaller base of investment the proportionate damage is about twice as great to the latter. In most of the major categories of public social investment, nevertheless, the record is creditable. Despite a rise from 7.6 to 9.1 percent in the defense share between 1965 and 1967, the total public education and health and hospitals expenditure shares went up from 4.5 to 5.3 and 1.17 to 1.23 respectively. And even federal spending for education and health, though not hospitals, rose. There are of course other costs involved in the inability to initiate needed programs—massive aid to the cities is the obvious example. But on maintaining or expanding established patterns of expenditure the score was not bad, at least until 1969 and 1970.

The pattern of federal expenditures for research and development indicates some recent but partially hidden costs to education and medicine. From 1955 through 1966 such expenditures rose spectacularly, and every year, from $3.3 billion to $14.9 billion.[23] Obviously such a skyrocketing growth could not continue indefinitely; not even most of the beneficiary scientists expected it to do so. Still, if one regresses R & D expenditures against GNP over the entire period 1955 to 1968, the percent of variance accounted for is a very high 93 and the "estimated" level for 1968 is $19.4 billion, instead of the actual level of $16.3 billion. The actual level of expenditures in fact fell below the estimated level for the first time in 1966—the first year since 1961 when the defense share of GNP showed any notable increase. R & D funds have now leveled off, at a time of rapidly rising prices.

This leads to a very important sense in which many of these cost estimates are substantially underestimated. The entire analysis has necessarily been done with expenditure data in current prices; that is, not adjusted for inflation. Since we have been dividing each expenditure category by GNP in current dollars that would not matter providing that price increases were uniform throughout the economy. But if prices increased faster in, say, education or health than did prices across the board, the real level of expenditure would be exaggerated. And as anyone who has paid a hospital bill or college tuition bill recently knows, some prices have increased faster than others. From 1950 through 1967 the cost of medical care, as registered in the consumer price index, rose by 86.2 percent. Thus despite Medicare and Medicaid, and even though the health and hospital share of public expenditure rose in current prices, the real share of national production bought by that spending *fell* slightly, from 1 percent to about .99 percent.[24] Comparable data on educational expenses are less easy to obtain, but we do know that the average tuition in private colleges and universities rose 39 percent and in public institutions 32 percent over the years 1957-67.[25] This too is faster than the cost of living increases over these years (not more than 20 percent), but not enough to wipe out a gain for government education expenditures in their share of real GNP.

In summary, past tradeoffs between defense and civilian needs indicate what some of the opportunity costs of defense spending have been. Some of the losses have been in consumer frivolities; many have been more damaging if not so immediately painful. It is important to recognize that major regularities *have* been found. The relationships between defense and civil spending are not so idiosyncratic or bound by circumstances that no general statement can be made. Perhaps many of the costs were unavoidable, to meet pressing defense demands that, if not supplied, would have damaged the long-run welfare of the country in other ways. But some of the costs were inadvertent, the result of choices whose implications were not fully understood. Thus even when the defense expenditures were fully justifiable in terms of most Americans' interests, the costs were not necessarily distributed in a way consonant with these interests. Where the defense spending was not necessary, the damage to the rest of the society is doubly regrettable.

6. The Costs of Defense in Other Countries

Defense vs. Civil Needs in Many Nations

The United States is of course not alone in having to choose between defense and other kinds of consumption or investment. Other governments, whether they make the choice consciously and with good information or without awareness of the options and the consequences of their actions, have to make the same choice. The choice is made just as much by a decision to let things take care of themselves and allow the market to distribute the burdens as it is by deliberate policy to soften the impact on investment or, by raising certain taxes, to have it absorbed particularly by certain kinds of consumption. We shall look briefly at the available evidence, which compares the experience of a large number of nations, then turn to what has occurred in Great Britain over a long seventy-five-year span, and finally compare the recent experience of three major Western nations, Britain and also Canada and France. It will be clear that the experiences of various nations differ quite a bit one from another and from the American pattern, strongly suggesting that at least in principle a good deal of choice is available. How a nation will in fact distribute the burdens of defense spending is fixed less by inexorable laws of economics than by the nation's political system and the values of its people.

The initial evidence points clearly toward this conclusion. After looking at a limited amount of data for some Asian nations, Charles Wolf, Jr., found no particular uniformities and decided that each political system absorbed the economic impact in a different way.[1] Similarly, Frederic Pryor examined the short-run effects of defense changes on various types of expenditures in most Western developed nations over the span of the 1950s and early 1960s. In those nations where defense spending was relatively high (except France) he did find an inverse relationship between defense and all public civil expenditures; in the small states where the defense proportion was low, however, no relationship appeared. Some relationships of defense to other GNP categories were found in individual countries, but there was no standard pattern.[2]

Similar results arise from cross-sectional comparisons of relative spending levels in different countries at the same point in time. Looking at Western European nations in 1962, Pryor found a notable (statistically significant) inverse relationship only between total investment and defense as shares of GNP. Using similar data for the year 1956 no relationship between defense and investment appeared, although there was some evidence then of a general pattern of substitution between defense and civil government (as we noted above, he found a similar relationship within several nations over time). But finally, using some carefully refined cross-sectional data on a special sample of fourteen market and centrally planned economies, he found *no* general relationship between defense and other major expenditure categories.

Much the same is true with the data on approximately 120 nations to be reported in the second edition of the *World Handbook of Political and Social Indicators.*[3] Correlations between defense on the one hand and gross domestic capital formation and public expenditures for health on the other, each as a proportion of GNP, showed none of the variance in either of the latter to be accounted for by defense. Essentially the same thing occurred when military manpower as a fraction of adult population was used as the attempted explanatory factor. Public expenditures for education were slightly positively associated with defense spending and military manpower shares, but the relationship is so low (11 percent of the variance accounted for in each case) that although statistically significant at the .001 level the finding does not seem very important. An earlier effort to relate military manpower and defense expenditure ratios to consumption and gross capital formation was similarly inconclusive. The only notable correlation showed military expenditures accounting for 17 percent of the variance in consumption as a percentage of GNP, with an inverse relationship (significant at the .002 level).[4]

This failure to find strong relationships between defense and civil expenditures is really not surprising when one considers the matter carefully. If defense spending does not take a very large fraction of a nation's resources, there are indeed many different ways in which the slight burden can be carried; different decisions may well be made by different countries or by the same nation at varying points in time. We could only expect to find notable regularities in nations where the level of defense expenditures is high enough often to force some painful choices. Since more than two-thirds of the world's nations spend less than 4 percent of their GNP on defense, comparing their experiences is not especially likely to be fruitful. Furthermore, when looking at a single nation over time we should have (1) a reasonably long time span so as to encompass periods both of heavy military spending and of relaxation, and ideally several of each; (2) a fairly high average level of expenditures; and (3) a wide enough range of variation to provide situations where the impact of military changes will be sufficiently severe to force hard choices or offer opportunities.

Military and Civilian Spending in Britain since 1890

By these criteria, the best data available are for the United Kingdom, covering the period 1890 to the middle 1960s. This is an appreciably longer span than can be examined for the United States, and, since it includes the turn-of-the-century naval race with Germany and very heavy military expenditures in World War I, it is in some ways an even more interesting case than the American one. What is more, Britain's balance of payments difficulties and strained economy of recent decades have meant that even the comparatively moderate defense exertions of the 1950s and 1960s have inflicted notable costs on the country. Figure 6.1 shows the time trend of British defense expenditures as a proportion of GNP. Note that the British did not suffer a "ratchet effect" after World War I, being able to reduce their proportionate military effort to its prewar level. The British political system, with power more centralized in the cabinet and/or party central offices, is less susceptible to legislative logrolling than is the American system, and this may account for the difference.

Unfortunately, but not surprisingly, data on the desired spending levels are rather sparse, and we are required to work with considerably fewer than were available for the United States. For this examination I have looked only at those expenditures where the data cover both world wars. Table 6.1 shows the results. Most striking is the negative relationship between defense and personal consumption, just as in the United States. To be

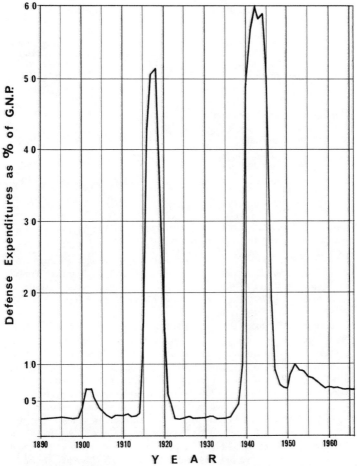

Figure 6.1 United Kingdom Defense Expenditures as a Percentage of GNP, 1890-1966
Source: See table 6.1.

Table 6.1. The Effect of Defense Spending on Civilian Activities in the United Kingdom, 1890-1966

	Percentage of Variance Explained (linear regression)	Regression Coefficient
Personal Consumption	51*	−.360
Exports	40*	−.164
Imports	05	−.069
National Gov't Civil Spending	01	−.032
Local Authorities Expenditures	01	−.022
Total Public Education	01	−.008
National Gov't Education	01	−.003
Local Authorities Education	02	−.008
Local Authorities Social Services	15*	−.019

*Both percentage of variance and regression coefficient are statistically significant at the .001 level.

Note: Not all years for all variables, though N always equals at least 53 and covers virtually the entire time span.

Sources: GNP: 1890-1953 from Alan T. Peacock and Jack Wiseman, *The Growth of Public Expenditure in the United Kingdom* (Princeton: Princeton University Press, 1961); 1954-66 from Central Statistical Office, *Annual Abstract of Statistics of the United Kingdom, 1961* (London: H. M. Stationery Office, 1961), and later issues. Defense, National Government Civil (includes transfer payments), and National Education, 1890-1936 from Brian R. Mitchell, *Abstract of British Historical Statistics* (Cambridge: Cambridge University Press, 1962); 1937-45 from Central Statistical Office, *Statistical Digest of the War* (London: H. M. Stationery Office, 1951); 1046-66 from various issues of *Annual Abstract of Statistics.* Defense figures for 1915 and 1920 were obtained by adding army and navy listed expenditures in Mitchell, p. 398, to votes of credit values on p. 399; 1916-19 figures are exclusively votes of credit. Defense equals army plus navy, plus air force after 1920. Exports and Imports 1890-1953 from *Historical Statistics;* 1954-66 from issues of *Annual Abstract.* Personal Consumption same except 1954 and 1955 also from *Historical Statistics.* Local Education 1894-37 from *Historical Statistics;* 1937-65 from *Annual Abstract.* Local Social Services same as GNP except 1954 from Peacock and Wiseman. Total Local Authorities from various issues of *Statistical Abstract of the United Kingdom* (predecessor to *Annual Abstract*). Includes all Ireland pre-1922 and Northern Ireland thereafter; 1919-21 figures for Ireland estimated.

sure, the percentage of variance explained by defense in Britain is not as high as in America, but it is still substantial, and the regression coefficient, showing how much consumption is lost with each dollar devoted to defense, is very nearly as high ($-.360$ as compared with $-.420$).

The other very strong relationship is between defense and exports, with 40 percent of the variation in the latter accounted for by the former. Furthermore, the regression coefficient is high, indicating a decline in exports equal to about one-sixth of any increment in defense expenditures, or, expressed the other way, exports typically go up by one-sixth the amount of any decline in defense share. The proportionate link to exports is even more striking than that to consumption and puts the two regression coefficients into proper perspective. Outside of wartime, British exports typically have run between 15 and 20 percent of GNP; consumption typically at around 80 percent. Thus the relative impact of defense spending on the export base is fully *twice* as great as on consumption. This points out a major part of Britain's postwar economic problem—a direct conflict between military purchases and export earnings, a conflict which is apparent not only in the aggregates but when one looks at the data on individual industries. Among Britain's most important actual or potential foreign-exchange-earning industries are aircraft, electronics, motor transport, and shipbuilding—precisely the civilian industries that divert major resources to military production when defense procurement rises. Some also engage heavily in arms production even in "normal" years, but often when the British defense demands are not too great they sell large amounts of armaments abroad, earning foreign exchange. So again, the conflict is very direct. For a country so dependent on foreign trade, the choice of arms vs. exports is cruel indeed and has played a major role in forcing a reduction of British defense spending in recent years, especially in that portion, such as supporting troops and bases in West Germany and east of Suez, where the immediate foreign exchange component was very high.[5]

As with the United States when we included years of total war, imports do not show the expected positive relation with defense, but the reason is again the disruption of normal trade by world war. Under 1946-66 conditions of "normal" trade and full employment, defense spending did put pressure on the import as well as the export side of Britain's balance of payments.

Local governments' expenditures for social services show a statistically significant but nevertheless not very strong negative relationship to defense. It is, however, much stronger for the post-World War II period, as we shall see below. For the other kinds of civil spending the variance explained by defense here is too low to warrant looking at the regression coefficients. Nevertheless, some relationships are apparent on examining the plots for the data, going beyond the simple linear regression model. For example, on each of the three education items, the second-order curve of defense explains about 15 percent of the variance. Education and defense show the expected inverse relationship in wartime years of major defense exertion (when defense is above 37 percent of GNP), but not in nonwar years (defense below 19 percent of GNP; no cases between 19 and 37). In fact, for all the nonwar years alone defense and education are positively related, as are defense and all national civil purchases. This latter relationship, however, is a result of a long-term secular increase in almost all British

central government expenditures, both military and civil
(a doubling in their GNP share since the 1930s), and no
causal effect of defense spending should be imputed. On
the contrary, we will show below that in the post-World
War II period defense and social welfare activities, including education, have been in direct and serious competition with one another. Also, the average level of
spending for education has been much higher in peacetime than in the war years. Between peace and war, the
regression coefficient is nearly −.07 for defense and
education expenditures. That is, in the transition from
war to peace a billion-dollar drop in military spending
ultimately was associated with an increase of almost $70
million in education.

Canada, France, and Britain since World War II

We can now turn to an examination of the post-World
War II experiences of the three developed Western nations that, with the United States, have both devoted
substantial shares of their economies to military efforts
and show notable year-to-year variation. Each of the
three has at some point since 1950 devoted more than 8
percent of its GNP to defense and is now below 6 percent. The data are shown in figure 6.2. The post-World
War II experiences of Canada and Britain have pretty
well paralleled each other, with the Canadian effort
somewhat smaller. Both have drastically reduced their
relative defense shares since the Korean War. France's
pattern is basically similar but complicated by special
efforts for colonial wars in Indochina and Algeria and
for the *force de frappe.* Data are not available on all the
expenditure categories for each of these countries, and
the time span covered varies a little, so we must be

cautious in pointing out differences among them or, especially, from the United States (which in the previous chapter included World War II data). Also, for all three countries there are far more years of declining defense expenditures than of increases; hence the regression coefficients will largely show what benefits when defense drops—unlike the American analysis, where defense increases and reductions were represented about equally. Some tentative conclusions can nevertheless be suggested. It will be clear that there is no single universal pattern for the impact of military spending.

In both Britain and France investment suffers some from high levels of arms acquisition and, especially, benefits from arms reduction. The regression coefficient is in each case quite high, three to four times as great as we observed in the United States ($-.29$). In France the absolute amount by which investment changed actually exceeded the variation in defense spending. Yet the high

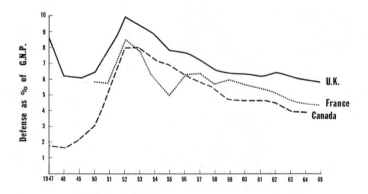

Figure 6.2 Defense Expenditures as a Percentage of GNP in Canada, France, and the United Kingdom, 1947-65
Source: See table 6.2.

regression coefficient is a bit less notable because the linkage between defense and investment is much less close; even in France less than half the variance in investment can be attributed to changes in defense (as contrasted with 72 percent in the U.S.). The reason is simple and important: typically French and British investment about held their own or even increased when defense was high, and often expanded strongly during military downturns. The result is a long-term growth in investment's share of both economies. But investment has not competed at all with defense in Canada; on the contrary there seems to be a mild positive relationship. At least at the level of exertion reached by Canadians over the past two decades, military expenditures may actually serve as a net stimulus to fixed capital formation.

Only in Canada does there seem to be a very direct conflict between defense and personal consumption. There the proportion of variance in consumption explained by defense is substantially lower than in the United States (84 percent), but the slope of the regression line is much steeper $(-.814$ vs. $-.420)$. It is impossible to tell from this kind of analysis whether the trade-off is the result of deliberate and explicit policy, but certainly the effect is strong. More than half of the additional dollar expenditures required for defense ultimately come out of personal consumption or, when the defense proportion declines, go back into consumption.

The French and British experiences are both quite different and more complex. The years of greatest defense effort in France were 1952 and 1953, times when the French economy still had not fully recovered from World War II, and there were some slack resources that could be mobilized by an infusion of public demand. In

Table 6.2. *The Effect of Defense Spending on Civilian Activities in Some Major Western Nations*

	Percentage of Variance Explained (linear regression)			Regression Coefficients		
	Canada (1947-64)	U.K. (1947-65)	France (1950-65)	Canada	U.K.	France
Personal Consumption	57*	06	35	-.814	.494	.825
Fixed Investment	25	30	49*	.529	-.842	-1.135
Gen'l Gov't Civil Consumption	16	48*	63*	-.290	-.469	-.440
Education & Research	01	35	49*	-.060	-.247	-.395
Health Services	01	05	24	-.031	-.064	-.041
Special Welfare Services	24	–	71*	-.180	–	-.027
Transfers for Ed. & Res.	–	40	–	–	-.065	–
Transfers for Health	–	00	–	–	.008	–
Social Security & Assistance	–	34	39	–	-.240	-1.569
Total Education & Research	–	37	–	–	-.312	–
Total Health	–	03	–	–	-.054	–
Total Welfare	–	–	39	–	–	-1.596

*Both regression coefficient and percentage of variance are statistically significant at the .001 level.

Note: For all three countries the defense percentages are appreciably lower than those used in the NATO study of chapter 4. For Britain and Canada, however, the difference is quite uniform over the years and would have little effect on the regression analysis. The difference between the two sets of French figures is pretty uniform from 1954 onward, but in 1951 and 1953 the figures show a slight decline in the defense share, while those of the NATO analysis show increases. I am not certain of the reasons, but here it seemed important to use figures from a source consistent with that for the other French public expenditure data.

Sources:

Canada: Dominion Bureau of Statistics, Information Services Division, *The Canada Yearbook 1950* (Ottawa: S. E. Dawson, 1950), and later annual editions; Dominion Bureau of Statistics, *National Accounts Income and Expenditure, 1926-1950* (Ottawa, 1951); United Nations, *Yearbook of National Accounts Statistics, 1957* (New York: United Nations, 1958), and *Yearbooks* for 1960 and *1966*; O.E.E.C., *Statistics of National Product and Expenditure, 1938, 1947 to 1952* (Paris: O.E.E.C., 1954).

U.K.: Central Statistical Office, *National Income and Expenditure, 1946-1952* (London: H. M. Stationery Office, 1953); Central Statistical Office, *Annual Abstract of Statistics, 1950* (London: H. M. Stationery Office, 1950); U.N., *Yearbook of National Accounts Statistics, 1957, 1966*. No data for 1954 on public expenditure breakdown. 1947 also missing for health services.

France: O.E.E.C., *Statistics of National Product and Expenditure;* U.N., *Yearbook of National Accounts Statistics, 1957, 1960, 1966;* Ministère des Finances et des Affaires Economiques, Institut National de Statistique et des Etudes Economiques, *Les Comptes de la Nation, 1949-1959* (Paris: Imprimerie Nationale, 1964); Ministère des Finances, *Annuaire Statistique de la France, 1954* (Paris: Imprimerie Nationale, 1954), and *1956, 1957, 1965, 1966;* Ministère des Finances, *Statistiques et Etudes Financières,* "Situation Provisoire des Recettes et des Despenses de L'Etat Pour 1949," (Paris: 1949), p. 260-62, also studies No. 23 (1950), No. 49 (January 1953), No. 52 (April 1953), No. 69 (September 1954), No. 73 (January 1955), No. 100 (April 1957), No. 142 (October 1960), No. 160 (April 1962), Nos. 163-64 (July-August 1962), No. 197 (May 1965), No. 220 (April 1967).

those years the goods and services needed for the military effort came in large part out of slack and not in any case much at the expense of consumption. Private consumption essentially maintained its fraction of GNP during the buildup and then fell after the first push of the defense effort was largely completed, accounting for the moderate positive association we observe between trends in defense spending and in private consumption. This drop in relative defense spending since the early 1950s, despite the French nuclear effort and *force de frappe,* has perhaps enabled investment to rise. Certainly it is capital formation rather than consumption that has in fact increased its share of resources over the period. In Britain, on the other hand, the level of private consumption has been quite unrelated to the defense effort. British consumers were forced to sacrifice for the rearmament of the early 1950s, but like the French they were not especially indulged when, in the late 1950s and 1960s, military spending was curtailed. In both countries the share of personal consumption has dropped to about 64 percent of GNP, from almost 69 percent in 1951.

For France and Britain it is quite apparent that defense expenditures have been negatively correlated with government spending for civilian needs. In both cases the former explains roughly half the variance in the latter, and the regression coefficients are remarkably similar. Each country shows a growth since 1950 (by more than 1 percent of GNP) in the civil government's share of the economy.

Actually, the figures in table 6.2 tend to understate the dependence of government spending on defense in the United Kingdom, because they apply only to the linear model. A curvilinear fit adds more than 25 percent of the remaining variance. The curve shows a steep downward slope (inverse relationship between defense and civil government) over most of the range but flattens out to a low level plateau for the greatest Korean War period rearmament exertions. The slope for the low-to-middle defense levels then becomes much sharper than appeared with just the linear regression model; for that range the coefficient becomes approximately -1.0. In other words, from the mid-1950s onward every decrease in defense expenditures made possible a fully equivalent increase in civil consumption expenditures in the public sector. Adding together the five sets of figures for education, health, and social security, we find that these social services grew from a low of about 10 percent of GNP in the middle 1950s to a high of nearly 14 percent in 1965.

From a familiarity with British politics it is apparent that in contemporary Britain social welfare expenditures have become the "hard" demands in the political system to which defense pressures must be accommodated, not the reverse.[6] In foreign policy the British have accepted an increasing dependence on the American deterrent as the price. Whether the long-run effect will be less or greater independence for the United Kingdom is not clear, but the fact that private consumption has not been indulged by the defense reductions suggests that some effort has been made to focus upon the ultimate social and economic bases of national strength.

The Canadian situation has again been different. The linear regression shows only a very small relation between military spending and public nonmilitary consumption. These figures are deceptive, however, because a curvilinear fit adds almost half the remaining variance. Government civil spending held its own pretty well during the Korean rearmament; since then it has grown almost as steeply as the defense budget has dropped.

Finally, we must look at the breakdowns for public expenditure. Some of the categories are subsets of government consumption, others come under the rubric known as "transfer payments." According to the standard accounting definition, transfers are unilateral payments from government that are considered to provide additions to the income of their recipients, who may be households or private nonprofit institutions. They are in effect subsidies rather than payments for goods or services received by government; they include pensions, unemployment compensation, public scholarships, and subventions to educational institutions. Except for social security and public assistance they are typically fairly small as compared with government consumption; in Britain, for instance, the health and education transfers amount to only one-fifth to one-fourth the amount expended under consumption for the same categories.

In Britain and France, expenditures and transfers for health are not much affected by changes in the armed forces budget. Most of the major components of these aggregates represent long-term social commitments that at least in the European countries are not readily changed. Welfare and/or social security assistance responds somewhat more strongly to changes in the defense budget, but, as in the United States, this is not especially surprising. Periods of heavy defense expenditures are likely to be characterized by low unemployment and the return of some retired workers to the active labor force.

The effects of defense needs on education, however, are more interesting. Expenditures for education and research suffered relatively little in periods of heavy military demand. But to put it in the direction that military spending has been going in those nations over the last decade and a half, the diminution in relative defense costs has made possible substantial increases in educational efforts. Since the early 1950s both these nations have pared their defense budgets by an amount that is equal to about 4 percent of their GNPs. This has been accompanied by a shift of over a third that amount (1.8 percent of GNP) in France into educational enterprises. Since public educational expenditures amount to only 3 or 4 percent of GNP in any case, the improvement has been very great when measured against the base from which it began. Coupled with the relative increase in investment over these same years that we noted earlier for France and Britain, it is plain that they have turned their defense savings to good long-term use, substantially expanding the productive resources of their economies.

At first glance defense in Canada appears unrelated to health and education budgets. The linear regression coefficients nevertheless hide a most important development in the postwar Dominion, and a curvilinear fit accounts for about 45 percent of the variance in each case. Since 1947 Canadian public health and education expenditures combined have risen from 3 percent of GNP to more than 8 percent. This expansion was halted, but not reversed, during the Korean War rearmament period. Since then it has proceeded under great steam. In fact, the decline from 8 percent to below 4 percent of Canadian resources devoted to defense since 1953 has been matched by a fully equivalent transfer of resources into the health and education sectors. With fixed capital formation we found that Canada did not earn high marks specifically for expanding investment with the funds released by savings in military spending—capital formation in recent years has taken just the same proportion of national income that it did in the early 1950s. But its social investments in a healthier and better-educated populace are impressive indeed.

A Comparative Evaluation

Normatively, it is difficult to make a clear-cut evaluation of the overall virtues and defects of each nation's experience. None of the countries examined has any monopoly of either. On the requirement that defense upswings be financed largely out of personal consumption, both the United States and Canada come out well over most of the period; on the requirement that capital formation and public sector social investment in education and health be insensitive to military demands, the United States performance is mixed. During the post-World War II and Korean War rearmament periods investment pretty effectively held its own in Canada and, to a lesser extent, in the United States. For example, in the early 1950s the Canadian fixed investment percentage actually rose slightly, and health and education combined fell by only .3 percent of GNP. In the most recent American military expansion, investment has suffered proportionately more than consumption, though the health and education shares have actually risen.

Another criterion, however, is that in periods of declining military needs the released resources be kept largely for investment and education rather than returned to private consumption. Here one finds that the United States has performed not too badly with its fixed capital formation. Some revival of investment occurred after World War II and the Korean War, and even during the modest 1962-65 dip in defense expenditures. But over the last twenty years any increases have been small, and the recent level of investment is well below that of the mid-1950s, although the defense share has dropped slightly. And since 1953 Canadian defense expenditures have been fully halved as a proportion of GNP without any systematic increase in investment to take advantage of the opportunity presented.

Britain and France both fit the ideal picture a bit better. The French have been particularly good at substituting capital formation for defense during the latter's downturns, though investment did slip during the worst of the Indochina war; the British were very good at maintaining investment during periods of expanding military demands, and since Korea have also taken good advantage of a declining arms budget. Both nations exhibit a long-term secular rise in investment between 1953 and 1965. In each case investment's share of the GNP has gone up by a third.

In Britain, civil government expenditure has expanded its share of GNP, over the prewar base, after major wars. Wars brought higher tax revenues and new pressures for public services; when military spending declined thereafter, public spending in large part replaced it.[7] But no such displacement effect occurs in the United States. Post-World War II federal purchases in total have never even approached their 1939 share of GNP; a lower peak in 1949, preceding the Korean War, also has not again been reached.

As for education and research funds, the United States was for many years fairly sensitive to defense requirements on both the upswings and the downturns of the latter. In the eyes of many American leaders educational needs apparently could be sacrificed or postponed when

military needs were great but were rather readily indulged when defense pressures relaxed. This may now
have changed. In the last decade education went up even
when defense did, as was earlier often true in Britain
and France. Most recently, however, American educational spending has been pinched by the continuing demands of war (for example, President Nixon's veto of a
high education and health appropriation in early 1970).
Nevertheless all three countries can point to a long-term
secular increase in the proportion of resources going to
education, a trend on which the fluctuations, with upward and downward slopes, are superimposed. The
Canadians have most dramatically devoted a rising share
of their resources to education in every year but four of
the last eighteen; this long-term growth in commitment
to education has brought its GNP percentage from 2.1
to 5.2, an upward trend broken seriously by military
needs only in 1951. In 1952 education resumed its
growth despite a concomitant rise in military expenditures.

The experience of other nations suggests some possible alternatives for American decision-makers. On the
whole, these other countries seem to have been a little
more successful in preventing military expenditures
from holding back attention to social or physical investment. Partly that is because ultimately their military
security now depends less on their own efforts than on
American protection, so they will not make very great
sacrifices for their armed forces. But it is probably no
coincidence that each of these countries (France perhaps only since 1958) has a strong executive, which
controls policy-making and has command over a disciplined legislative majority. Particularistic pressures can
more easily be resisted, and temporarily unpopular anti-

consumption programs somewhat more easily maintained, than in the American system with an independent legislature and weakly disciplined parties. That may in large part also be why these countries have actually been able to reduce their relative defense expenditures so notably in recent years. We earlier remarked on the absence of a "ratchet" effect in Britain and attributed it to this.

Past opportunity costs in the United States do not provide a perfect guide or deterministic mold for the future. Both the political system and the economy change. Should it be concluded, after enlightened discussion, that certain new defense needs must be met, it is possible by careful choice and control to distribute the burdens somewhat differently. If costs cannot be avoided, perhaps they can be borne in such a way as better to protect the next generation. At the same time, we must recognize that there are many features of great stability and resistance in American politics. The pressures that have in the past directed the costs of defense toward investment and government civilian programs are likely to retain most of their influence. The American political system is composed of a different balance of pressures than are those of Canada, Britain, and France, and the values of Americans are not precisely the same as those of our allies. Any decision to maintain or increase military spending, therefore, will probably have effects not substantially different from those in the past unless a very great effort is made to do things differently. It may well be as easy to vary the level of military spending as to change drastically the distribution of tradeoffs. If so, careful evaluation of military demands is all the more essential.

7. The Military in American Life

Diagnosis

Military expenditures in the United States are high; for the last twenty-five years they have regularly taken a greater share of this nation's produce than at any time in its history, other than in periods of all-out war. In large part the burdens of a big defense budget have been thrust onto America by the demands of the international system, by some mixture of Soviet aggressiveness and the arms race pressures inherent in a bipolar world. Perhaps too, American belligerence (or is it a basic insecurity?) has played a part.

It is impossible to sort out the relative weight of individual influences in the international system—whether one side has been aggressive and the other merely reactive, or whether a mutually reinforcing process of fear and response has been at work. Nor can we say with confidence how important *all* the international influences are relative to pressures in the domestic political system. But it does seem very likely that such domestic pressures also play a role, inflating the American defense budget beyond the size necessary to meet plausible threats from the Soviet Union. Each branch of the armed services plays bureaucratic politics, looking out for its own interests and the nation's as it sees them. In so doing it makes common cause with many congressmen, and logrolling in the legislature, along established patterns that apply to many civilian programs as well, helps keep military expenditures high. America's alliances have been of questionable value in substituting foreign military strength for an American armed presence. While there are a few exceptions, most of our allies spend a much lower proportion of their resources on defense than we do. For them the alliance presents an opportunity to rely primarily on a strong protector; for the United States, all too often the alliance is an increment of commitment without a corresponding increment of military strength.

If defense spending is higher than it needs to be for the maintenance of external security there are serious costs to the domestic political and social system. One obvious set of costs is the neglect of socially beneficial programs because enough money is not available: investment and public expenditures for health and education suffer. New programs to meet desperate social needs are even harder to fund than old ones. Other countries have had the same experience and, most notably Britain, have reduced their military programs accordingly. In the long run, a nation's strength depends on continuing its investments; present arms can be bought at the expense of future strength—even future military strength!

An equally pernicious and less well-known cost arises in distortion of the political system. The pattern of congressional voting on Defense Department appropriations is very similar to that of voting on a wide variety of issues related more generally to defense, arms control, and East-West relations. Congressmen line up much the same way on all these issues, and their alignments are related, in no insignificant way, to the economic importance of Department of Defense payrolls and employment to their constituencies. The larger the DoD share of employment in their states, the more "conservative" they are likely to be. This applies also to senators' positions on gun control and is true regardless of party or region of the country. While cause cannot be proven in any strict sense, it seems very likely that maintenance of a large defense establishment helps to sustain the hawk-ish and most uncompromisingly anticommunist forces in American political life, with a viewpoint stressing excessive "preparedness," a hard-line foreign policy, and overreliance on military power in our dealings with other nations. Without surrender, ultimate control of the arms race and the flowering of American politics both require the mitigation of these attitudes. The size of the military is not just a matter of a little waste, unfortunate but tolerable. Surprisingly, however, con-gressional attitudes on arms issues seem unrelated to contract awards, indicating that the political weight of the industrial half of the military-industrial complex is less conservative than has often been thought. Not all the alleged offenses of defense are real ones.

The United States probably does not need to fear a coup by its soldiers, on the order of *Seven Days in May,* the French army's attempt in 1958, or the politics of Latin America or other underdeveloped areas. At least in established political systems, military coups require a tradition of military involvement in politics that is utter-ly lacking in the United States. Few Americans expect their soldiers to take direct political action, and the army itself is well socialized to the norms of civilian primacy. Though the basic ideology of most military men is rather conservative, it is rarely fascist and rarely leads to overt acts of political insubordination. Military men essentially view themselves as legitimate in politics only for the purpose of maintaining their role as defend-ers of the nation from external danger. The General Walkers are striking because of their rarity. Maintaining good relations with congressmen and in the civilian ex-ecutive branch is one thing; direct politicking in the public arena, let alone revolt, is quite another. And protecting the national security also is narrowly conceived —America's military officers do not see themselves as needed to modernize the country, to bring political order, or to eliminate civil corruption, as often their counterparts do in the emerging nations.[1] Furthermore, there are so many routes to personal wealth, power, and prestige in the United States that a career pattern of general-to-dictator has little seductive potential.

It is possible that severe and repeated military defeat could widen officers' conceptions of their role in protecting national security. A pattern of what seemed civilian incompetence in the face of conflict could make soldiers feel the country's interest required them to become politically more active. This risk is, I suppose, in some degree inherent in Vietnam withdrawal efforts, but it would probably take more than one such defeat to change the climate seriously. So long as the normal political order remains reasonably efficacious and holds the confidence of most citizens, military men are unlikely to be tempted to reach for wider power. A further breakdown in the political system might make the reins of power more inviting, but despite America's agony that danger does not yet seem close.

The real and present problem with the armed forces in American politics is how to control spending for armaments, and how to limit the spillover from alignments on defense expenditures into other issues of foreign and domestic policy. We must worry not about a sudden take-over of power by our soldiers but about how to prevent slow accretions in the scope of military influence in the "normal" political system. Here we must concern ourselves at least as much with the military's civilian allies, who use arms spending for their own purposes, as with our soldiers and sailors. We need to con-

trol military spending because we cannot afford to
waste material resources and because excessive arms
expenditures distort our political and social life.

We also need to limit arms spending because of the
dangers of an arms race. So far, the arms race has not
been rapidly run. While the absolute expenditure levels
have risen, neither the United States nor the Soviet
Union now devotes a larger share of its national resources to arms than it did in the mid-1950s. Were it not
for Vietnam, the United States would show a clear decline in its defense share since the early 1960s. Also, the
technology of the last decade has been kind to us, Russians and Americans alike—more kind than most of us
realize in the face of awful weapons of death and pain.
The kindness has been in the ultimate strength given to
the defender. In a world of balance of terror, both
superpowers have been able to build secure deterrents,
invulnerable to surprise attack. So long as that invulnerability lasts, the initiation of nuclear war will remain
unattractive.

The invulnerability of our deterrents, however, is not
likely to be a permanent gift. "These things too shall
pass away," and it is myopic not to use the surcease to
develop other, safer, cheaper, and more permanent
means of preventing war. And technology is not au-
tonomous; the hasty, ill-considered, or foolish acquisition of new weapons could rapidly erode the invulnerability of one side or both. ABM systems, MIRVs, orbital
bombs, and myriad other possibilities *could* destroy the
contemporary balance that preserves each side's second-
strike capability. This is the greatest risk of the arms
race; that we and the Soviets will do whatever is tech-
nologically feasible without properly considering its military and political implications. Like climbing moun-

tains, trying to build new weapons just to see if it can be done has its attractions. The attractions are enormously reinforced by pressures of domestic politics and by Soviet and American fears of each other.

At the same time, we must not forget that a failure of deterrence can also come from a *neglect* of our weapons, both current and projected. Until there is an essential change in the international system, a failure to buy a needed new weapon could have effects just as calamitous as a runaway arms race. The country needs a mixture of carefully chosen weapons, both conventional and nuclear, to meet a variety of possible threats. Lack of conventional weapons might force us to resort to the nuclear option too quickly, as in the case of a moderate outbreak of conflict in Europe; overreliance on a single nuclear delivery system would leave us vulnerable to technological breakthroughs in our enemy's defense against that system. For these reasons too, money wasted on the wrong wars or the wrong weapons threatens our security, because after the waste we then might not have the resources for what we needed. For these reasons too, sound evaluation and control of military spending is required. Not even the most ardent advocate of preparedness can favor a system that blinds rational choice, retains old programs beyond their usefulness, and selects new ones according to the accidents of political influence.

Maybe the greatest risk from possessing a needlessly big military force is the temptation to use it too readily. In the contemporary world, the United States is the strongest nation and, whatever its faults, remains the protector of individual liberty in many states. In its hands, military force becomes Tolkien's One Ring of Power. On occasion we must wield that power to defend ourselves and our friends and to keep the Ring from passing to our enemies. For example, I think its use in Korea was appropriate and restrained. Yet employment of the Ring must be rare and restricted to cases of great necessity. Used rashly, unworthily, or even often, it will corrupt its bearer. Perhaps the United States, by its history and its ideals, carries some limited degree of immunity to the Ring's curse. But excessive reliance on force will quickly weaken, not strengthen us, and ultimately we will be no better than those we oppose.

If we are to establish and maintain enlightened democratic control over defense expenditures, it is essential to understand both their causes and their consequences. If we think that heavy military burdens are thrust upon us *solely* by the forces of international conflict, we will never be able to evaluate properly the new military requests that will be made of American taxpayers. Nor if we fail to understand the consequences of such expenditures will we have the incentive to limit them.

Prognosis

At the same time, we must recognize that diagnosis is easier than prescription. The disease is not one easily cured. Military expenditures have grown in the United States and have rarely been succesfully challenged because of the absence of any strong political or institutional base for an antimilitarist "peace lobby." More generally, Samuel Huntington has noted the lack of strong central political institutions in the United States as compared with the European democracies.[2] This country was founded in opposition to a central authority by men who derived their political theory and built their new institutions according to a design inherited from the opponents of Tudor absolutism. They constructed institutions for a separation of powers, both between central and local governments and among the branches of any one governmental level. As a result, it is very hard to mobilize the entire political system for social, economic, or political change of any kind, even kinds desired by a majority of the populace. "Incremental" change is a more apt description of the output of American politics than of most other nations.

For civil-military relations, this means there is a strong degree of protection against widespread, pervasive military domination of the nation. But it also means that once an interest, in this case the joint interest of the military and its civilian allies in maintaining heavy defense spending, is firmly established in the political system, it is extraordinarily difficult for others to cut back that interest's power. The promilitary group cannot pervert the whole system, but its members can use the system extremely effectively in pursuit of limited goals—goals that nevertheless put enormous hazards in the path of other goals sought by other members of the system. The lack of strong central political institutions in the United States makes it hard first to restrain the military and its civilian allies in their spending and secondly, in the absence of sufficient restraint, to apportion "properly" the costs of military spending. Lacking truly strong central direction, executive and especially legislative policy-makers find it easier to cut physical and social investment than to reduce current consumption. Future generations suffer because their political lobby is weak.

For these reasons, I very much doubt that the answer to bureaucratic politics and executive approval of military spending is to concentrate more power over defense matters in the hands of the Congress. We cannot control the military by giving power *to* Congress unless we first concentrate power *in* Congress, building under those with decision-making power a power base that will enable and encourage them to resist defense demands. Though many legislators now are becoming opposed to heavy military spending, in the 1950s Congress typically urged upon the executive bigger defense programs than the President or his cabinet wanted. That pattern is not entirely dead even now. Many congressmen now hate the Vietnam War, but the majority of congressmen quite clearly are not yet prepared to put an end to it. Many congressmen opposed the ABM and indeed almost defeated it in 1969. Yet they did not defeat it. For almost a decade previously the ABM was held back by technical, strategic, and bugetary arguments by forces within the executive branch. If the decision had been up to Congress, would it have done so well?

Power in Congress is too dispersed, in the sense that congressmen ultimately depend for political survival on the approval of their individual constituencies rather than on a central party organization or a national constituency. Most of their campaign funds come from constituents; if they are defeated they cannot easily move to another district. There is little basis for firm party discipline. Congressmen therefore must respond closely to the demands of their constituents and must show in sharply visible ways that they can look out for their districts' well-being. Most legislators also establish their own politically profitable relationships with lobbyists or at points in the federal bureaucracy. So long as this is the case, congressmen will find it hard to resist the political pressures from defense-dependent interests. The degeneration of tax reform into a tax cut in the 1969

Congress is an example of the inability of congressmen to resist the demands of particular interests and the power of logrolling to obtain the acquiescence of other legislators even when the general welfare is not served. Yet another example is the long-term history of restrictions on international trade. The beneficiaries of any particular import restrictions are concentrated into a few districts where they exert great influence on their representatives; consumers are the beneficiaries of trade liberalization, and they are diffused, unorganized, and only indirectly affected. Thus, without a change in congressmen's power bases or greater party discipline imposed by men less vulnerable to particularistic defense pressures, giving greater power to Congress as a whole would be unlikely to improve matters much.

Effective limitation of military spending depends in some part on weakening the power of constituency promilitary interests. A generalized "peace lobby" may help some in mobilizing countervailing forces in districts where the military impact is slight. Another help may come from the apparent fact, from chapter 3, that the political impact of DoD contract expenditures is less severe than is the impact of DoD direct employment; hence any reduction in military spending may have double political benefits if it emphasizes cuts in the number of men in the armed forces. More important than anything else, however, is probably the emergence of powerful demands for alternative spending and their establishment of "constituencies" in the political process. Demands for better health, more education, better cities, pollution control, the fight against hunger are putting increasing pressure on government resources. It is now very clear, if it was not before, that federal spending is not infinitely expansible. Choices have to be made, and money for defense does come at the expense of other, desperately needed expenditures. Increasingly, defense will have to fight for its share of our wealth and justify its demands more carefully than before. In this way some good may even come from the recent massive cut in federal taxes, to be felt more deeply as the 1970s continue. Because there will be less federal revenue, in the face of mounting domestic needs military requests will surely be more closely scrutinized.

Some observers have urged that instead of adding up the sum of various components of a defense program to get to a total military budget, the total itself should be fixed at some level below that currently accepted and the individual services forced to bargain within and between themselves as to how the amount should be divided. In addition, a much lower manpower ceiling might be fixed for the armed services. Such a system, it is thought, would prevent the accumulation of demands against a weak executive or Congress and produce a defense establishment whose individual parts reflected military estimates of need rather than the vagaries of political influence. That system was perhaps approximated in the 1950s before the advent of the Pentagon systems analysts. However attractive, this seems to me to be a dangerous abrogation of civilian responsibility; even more than now it would divorce weapons selection from the choice of political goals for which weapons might be wanted. We might find ourselves buying what the technicians thought feasible rather than what we had any need for, or buying without considering the implications for arms control or strategic stability. Still, no method of military budget-making is perfect, and that one might at least succeed in enforcing a lower ceiling while maintaining incentives to guard against the most serious defaults. On the basis of past behavior there is some reason to think the Soviets would in response also cut back their total military spending. If they did not, American intelligence should be able to detect the fact, and most of us would then acquiesce in lifting the ceiling.

Typically in historic societies only a minority has had material ease; in the United States the majority lives well and a minority lives in poverty. Blacks, Spanish-speaking citizens, and some whites from poor areas suffer, but the majority is comfortable and not discontented. Even defense spending meets the classical criterion of logrolling by giving the appearance of offering something for almost everyone. Because the political system is built to resist change, it becomes extraordinarily difficult to correct the situation. It is perhaps even harder because in most cases inequality does not result from malign will. It is a consequence of the incremental growth of subsidies, tax benefits, public spending programs, and private actions, often initiated for good purposes, that have now become embedded in the system and perpetuate benefits for some members long after their broader utility has past. And it is ironic that because the majority is large the cost of surrendering some privileges would not in fact be great. For instance, to equalize the average income of whites and blacks would require a sacrifice of only about 4 percent of all whites' income.[3]

Trying to change long-established practices is among the most frustrating of political tasks; it is easy to conclude that to do so the system itself must be smashed. Yet a survey of other human experience provides little reason for certainty that a radically different system would be better. Perhaps I say this merely because I am

part of that comfortable majority, but I like to think
that my belief has more wholesome roots. Surely now,
with the popularity of concern for the environment
and the enormous damage threatened by unanticipated
effects of technological innovation, we must be cautious
about making drastic physical *or* social changes we do
not understand and whose consequences we cannot at
all foresee. I am skeptical of men's ability to create a
better world *just* because they have the intention to do
so. We need a new Utopia, not in the sense of a vision-
ary ideal but as a detailed blueprint of a better society,
combining the three goals of the French revolution with
far more success than was achieved by the activists of
1789. Like "hard-nosed" political science, political
philosophy must become "relevant." Lacking the blueprint, most Americans will still choose to reform the
system, perhaps even without great confidence in the
outcome.

Overall, we cannot delude ourselves that it will be
easy to make major revisions in any element of the
American political and economic system that is as
deeply entrenched as the defense establishment has
come to be. The traditional American polity is not
well; military problems constitute a major part of the
illness. The nation is probably approaching a crisis in
the medical sense, as pneumonia cases once typically
exhibited—the malady peaks and slow recovery begins,
or the patient dies. Either can happen. We are in the
pre-antibiotic phase of social treatment and have no
wonder drugs. To do nothing would be inconceivably
irresponsible. To act rashly and ignorantly would be
precisely the equivalent of leeching or bleeding, debili-
tating rather than strengthening the patient for the
crisis. We must try to act, and before acting try to un-

derstand. Despite all the imperfections and irrationalities of American politics, I do not believe it is naïve to think that knowledge is the beginning of power.

Appendix: Mathematical Models of Arms Races

by Peter A. Busch

Evaluating Arms Race Models: Criteria

An understanding of the general processes of arms races would obviously be helpful in explaining United States military spending. One type of research into the nature of arms races is the construction of mathematical models of competitive military spending. This review summarizes and criticizes a number of these studies in an attempt to determine what can be learned from them.

Clearly any evaluation of this literature depends upon what one expects of it. The most obvious demand that could be made is that we be offered empirically tested models that aid in the prediction of national military expenditures. Along these lines a frequent criticism is that the quantitative studies in this area founder on the basic inadequacy of the available data, especially for the contemporary period.

One such problem is the unreliability of the published statistics of many countries. As such scholars as Brody and Vesecky[1] have indicated, there is good reason to believe that Soviet military budgets diverge from actual expenditures when the Kremlin wants to emphasize its pacific intentions or its strong position but also desires to leave its strategic plans undisturbed. Chinese figures suffer from the same manipulations. While unreliable

I would like to thank Gary Brewer, Jeffrey Milstein, and Bruce Russett of Yale University for their criticism of various drafts. Gratitude is also due Peter Tan and Augustine Tan of the University of Singapore for their extensive mathematical criticisms and to Ruth Busch for her editorial assistance. All failings are of course my own.

data are a problem for any quantitative study, non-random errors can present insuperable difficulties—especially when the errors are systematically related to the political-strategic phenomena under study.

Another obstacle is the quantity of data. Any attempt to describe an arms race mathematically must consider the possibility that the proposed equations become irrelevant as the world changes. Not only may the various parameters not remain constant, the basic relationships may be altered by major events. Clearly each world war changed not only the composition of the "great power" category but also the nature of the interactions among the major states. This implies that since military budgets are reported annually the number of observations for any one historical period is likely to be too small to test a model that is sufficiently complex to capture the dynamics of an arms race.

If this viewpoint is tentatively accepted, then two conclusions follow. First, quantitative models of arms races cannot be tested with anything approaching desirable rigor. Secondly, this analytical approach is not now in a position to offer accurate, quantitative predictions of military spending.[2] Because of this, the present review takes hard-nosed empirical criticisms to be, in a sense, irrelevant to discerning the possible utility of such research.

What then is the use of such models? The contribution to be expected—indeed, demanded—from such models is that they organize and clarify the verbal concepts and theories which have been used to describe arms races and that they explicate the often unsuspected implications of our commonly used theories. The fulfillment of these criteria requires, that every term in the various equations have a meaningful political definition.[3] Equally important, every specified functional relationship must be politically plausible. Finally, the mathematical implications of the equations must correspond with reality. If these criteria are not met, then there is little point in worrying about data fits. Though some models can be rejected on empirical grounds, usually the data are so poor as to be compatible with a large number of mutually contradictory systems. A high correlation coefficient can inspire confidence in such a situation only if the model is substantively relevant and intuitively appealing.

Two overlapping classes of models are reviewed. The first category derives from the tradition of Lewis F. Richardson,[4] and has been called "interactive determinism"[5] for reasons that will be presented. The second might loosely be called "probabilistic strategic modeling" and is exemplified by the work of Arthur Lee Burns and Martin C. McGuire.[6] These types have been chosen because they are, at least in principle, amenable to testing with actual budgetary data. If we succeed in showing that these approaches are to be judged more by their qualitative clarifications than by their quantitative predictions, then these conclusions should be equally true of such analyses as game theoretic studies, which are further removed from the realm of military budgets.

Richardson and the Richardson Tradition

The pioneer in mathematical theorizing about arms races is Lewis F. Richardson,[7] who proposed that the complex issues of competitive military spending could be simplified and their essential core formalized into mathematical models. The assumptions to which Richardson reduced his topic have been neatly summarized by Thomas L. Saaty:

1. In an armament race between two countries, each country would attempt to increase its armament proportionately to the size of the armament of the other.
2. Economics is a constraint on armament that tends to diminish the rate of armament by an amount proportional to the size of the existing forces.
3. A nation would build arms, guided by ambition, grievances, and hostilities, even if another nation posed no threat to it.[8]

In mathematical terms, this two-nation (or two-bloc) model can be written:[9]

(1) $dx/dt = ky - ax + g$
(2) $dy/dt = lx - by + h$

Here, equation 1 denotes the increase in arms level over time for side X while equation 2 denotes the same for side Y. "x" and "y" are the respective absolute levels of armament while "dx/dt" and "dy/dt" are the corresponding rates of increase.

Consonant with assumption 1, the "defense coefficients" "k" and "l" signify the degree to which a nation is stimulated by its opponent's force level to increase its own military strength. "a" and "b" are the "fatigue" or "expense" coefficients and indicate the extent to which the burden of a nation's military expenditures inhibit further increases in spending. Finally "g" and "h" are the "grievance" terms that measure the impact of those circumstances that would inhibit a nation from disarming even if its opponent did so completely. Included would be historical antipathies, ideologies, the desire for conquest, and the need to have a certain number of troops for the maintenance of domestic order.

Perhaps the most obvious and most often commented upon aspect of this kind of model is that it is "deterministic."[10] These equations do *not* state that the relevant set of circumstances narrows the range of choices open to a government and that the leaders' actions can be predicted from these circumstances only in a probabilistic manner. Rather they state that a precise, "inevitable" relationship exists between increases in military spending and their assumed causes.

There are two aspects to this issue. First, Richardson believed that in many cases arms races and wars happen not because the participants want them but because leaders sometimes react to their environment in an unthinking, mechanical way.[11] However, if Richardson's formulations are viewed as regression equations then they are probabilistic in the sense that odds could be specified that decision-makers would act in the stated manner. As we shall see from the second kind of approach reviewed in this essay, a more realistic view of the world seems to result when the concept of probability is explicitly included in the model.

Two other kinds of criticisms can be made. The first involves the empirical testing of the system, while the second and more important query concerns whether the model fulfills the criteria suggested at the beginning of this paper. We shall deal with the empirical problem first because Richardson's difficulties with data closely parallel the obstacles faced by those who have built upon his work. We will then turn to the more serious matter of plausibility.

In testing his system Richardson first assumes that the defense and cost parameters for each side are respectively equal. He then combines the two equations yielding:[12]

(3) $d(x + y)/dt = (g + h) + (k - a)(x + y)$

To estimate the constants and to see whether the data

do indeed fit such a straight line, Richardson regresses $d(x + y)/dt$ (or rather the discrete form, $\Delta(x + y)/\Delta t$) on $(x + y)$ using four observations between the years 1909 and 1913.[13]

A host of problems are raised by this procedure. Most obviously, it is not very difficult to fit a line to four points. In addition, the regression has provided the parametric values not for equation 3 with its four constants but for an equation having only two constants, which might be written:

(4) $\Delta(x + y)/\Delta t = s + m(x + y)$

where $s = g + h$ and $m = k - a$. Here, as in the work of several other researchers, the deductive system and the regression equation used to test it are not very similar.

In order to deduce the values of "g", "h", "k", and "a", doubtful outside information and tenuous assump-

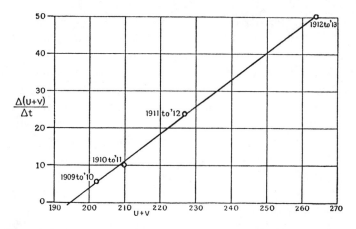

Figure A.1

Note: "U" is the empirical form of "x" and "V" is the empirical form of "y".

Source: Richardson, *Arms and Insecurity,* p. 33.

tions are used. A brief example will suffice to show the magnitude of the problem.[14] Although equation 3 has been tested with pre-World War I data, Richardson proposes to turn to the post-1918 period to measure "g" on the implicit and highly dubious assumption that the value remained constant. Following Richardson, let us take equation 1 to represent the actions of Germany between 1933 and 1936 and let y represent the combined arms level of Britain and France during the same period. Since the latter force level was fairly constant, set $y = y_1$ which is a constant. Finally, for the sake of simplicity, let us set $g = 0$ (despite the fact that g is assumed to be greater than zero in the rest of the analysis).

Then equation 1 reduces to

(5) $dx/dt = ky_1 - ax$

In 1933, Germany's weapons stockpiles were very small, so that we may let $x = O$. Then

$dx/dt = ky_1$

and

(6) $1/k = y_1 / (dx/dt)$

"$1/k$" may now be interpreted as the "apparent catching up time in an arms race from zero with no grievances."[15] Since Germany caught up in about three years, $1/k = 3$ years or $k = 0.3$ years^{-1}. In equations 3 and 4, however, "x" and "y" refer to blocs rather than to individual nations. Therefore, Richardson multiplies his "k" for Germany by a factor of two, since the pre-World War I Central Powers together were about twice as populated, well-armed, etc. as Germany taken alone.

The absurdity of all this requires little further amplification except perhaps to mention a point raised by Rapoport,[16] which serves to remind us that a lack of historical accuracy can vitiate quantitative manipulations as surely as can technical blunders. Specifically, the rationale for holding "y" constant in the estimation of "k" was that British and French arms levels remained constant. But as Rapoport suggests, the Soviet Union was clearly a potential enemy of Germany at the time, and Russian military spending was *increasing*.

As we have stressed, however, such aspects of empirical testing are not the most crucial matters to be raised in an evaluation of the services which arms race models can perform. Much more important is the degree to which such systems contain concepts and relationships that both embody and clarify our best qualitative understanding of competitive armament buildups.

An example of this type of issue is suggested by a comparison of the original Richardson equations, equations 1 and 2, and the relationship which is applied to the real world, equation 3. Over and above the problems already noted, it can be seen that equation 3 represents the combined actions of all relevant actors rather than the individual responses of each state. This analytical aggregation directs the theory away from distinctive national styles toward a focus on the world system. In effect, Richardson has said that arms races spring from worldwide causes rather than from the interactions of peculiar national or bloc characteristics. But the choice of this level of analysis demands a good deal of intensive, explicit, and intuitively plausible justification rather than the act of faith to which Richardson limits himself.

Given this focus, then, how plausible is Richardson's system? One way of answering this question is to deduce and interpret the mathematical implications of the equations and to compare them with reality. Particularly interesting in this regard are the formal "stability" and "equilibrium" conditions of the model.[17] The concept of "equilibrium" refers to the point at which the rates of growth, dx/dt and dy/dt, are zero. It is important because it allows us to ask what circumstances will, according to our model, lead to a halt in the arms race.

The "stability" conditions are those circumstances under which the pause in military stockpiling will last. If domestic or international events cause a resurgence in the armament race, and if the equilibrium of the system is a stable one, then both sides will tend to move back toward the equilibrium point once the disturbing influences are removed. If the equilibrium is unstable, then any "displacement" from the equilibrium will result in a continuing arms race. Of course the model may imply that no point of equilibrium exists. In summary, not only do the original equations specify how competitive arms buildups proceed, but they imply the conditions under which the process might stop.

Since William Caspary[18] has very lucidly examined this matter in the Richardson model we shall closely follow his argument here. If we examine equations 1 and 2 it is clear that the equilibrium is that point where the force levels do not grow, i.e. where the derivatives equal zero. In this case we have

(7) $dx/dt = 0 = ky - ax + g$

(8) $dy/dt = 0 = lx - by + h$

Relabeling the constant x and y as x_0 and y_0 we may solve equations 7 and 8 yielding

(9) $x_0 = (hk + gb / (ab - kl)$

(10) $y_0 = (gl + ha / (ab - kl)$

Clearly the position of x_0, y_0 depends on the values of the parameters. Let us assume that all the constants are positive. Then figure A.2 shows the case for ab − kl>0 or ab>kl while figure A.3 shows the equations for ab<kl. In addition the diagrams show the signs of the derivatives for different positions on the x,y plane. The latter can be derived by noting the effect, for instance, of varying x in equation 7 while keeping y constant. Thus, above the line dx/dt = 0, dx/dt is positive and below it the derivative is negative.

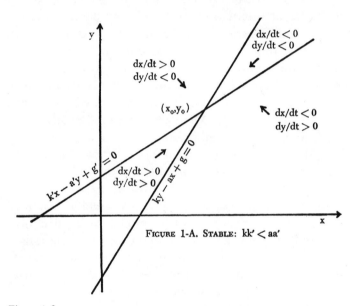

FIGURE 1-A. STABLE: $kk' < aa'$

Figure A.2
Source: Reprinted from Caspary, "Richardson's Model of Arms Races," p. 66, by permission of the Wayne State University Press.

If, in figure A.2, there is a displacement away from the equilibrium because of factors not included in the equations, then the arms levels of both sides will move back to x_0, y_0 as indicated by the signs of the derivatives. If, however, ab<kl as pictured in figure A.3, then the equilibrium position is in the negative portion of the

x,y plane. But this is not possible since negative arms levels are politically meaningless. Thus no stable equilibrium can be reached if an arms race follows the processes stipulated by the latter parametric values in the manner prescribed by the equations. A temporary pause in the arms race might result from influences not explicitly included in the equations, but the model suggests that unlimited escalation must soon follow.

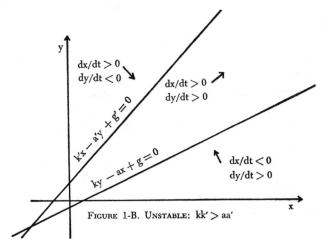

FIGURE 1-B. UNSTABLE: kk′ > aa′

Figure A.3
Source: Reprinted from Caspary, "Richardson's Model of Arms Races," p. 66, by permission of the Wayne State University Press.

In substantive terms, figure A.2 may be summarized as claiming that "beyond the equilibrium point the burden of armaments in the system is greater than the incentive to arm. Below the equilibrium point the reverse is true."[19] Furthermore, the stronger the grievances of each side, and hence the larger the values of "g" and "h", the higher the arms levels at the point of equilibrium. In contrast, figure A.3 shows the case where the incentive to arm is always stronger than the economic burden of weapons.

It would seem from this discussion that the equilibrium and stability conditions implied by the Richardson model are quite reasonable. But if we return to equations 1 and 2 we note that the incentive to increase one's armament level is proportional to the *absolute* level of one's opponent's forces. However, it is clear that what causes a nation's leaders to increase their armed might is the *disparity* between their military strength and that of their foes.

Richardson attempted to come to grips with this issue by formulating a "rivalry" model. Instead of multiplying the reaction coefficient of one side by the arms level of the other (ky and lx) as in equations 1 and 2, Richardson multiplied the respective coefficients by the difference between the two arms levels (k $[y - x]$ and l $[x - y]$):[20]

(11) $dx/dt = k(y - x) - ax + g$

(12) $dy/dt = l(x - y) - by + h$

Rearranging terms, we get

$$dx/dt = ky - (k + a)x + g$$

$$dy/dt = lx - (l + b)y + h$$

And, finally, substituting into the original stability condition kl<ab we have

$$kl < (k + a)(l + b)$$

$$kl < kl + bk + al + ab$$

(13) $0 < al + bk + ab$

Since all of the constants in 13 are positive, the rivalry model states that a point of stable equilibrium always exists regardless of how aggressive, fearful, and heedless of costs the various nations are! This would be a comforting conclusion these days, but it is not true.[21]

The Pitfalls of Irrelevant Modeling

The debt owed to Richardson by all of the studies we shall examine is very large; even his mistakes have proven valuable by serving as stimuli for other studies. But this is true only because Richardson's equations were, if not correct, at least relevant to the subject. The form of the equations was chosen because they represented important influences in military competitions. Because of this an examination of his system was required in order to evaluate the model.

However, if the initial formulations do not meet the relevance criteria we have suggested there is no point in proceeding to stability conditions, data fits, etc. This issue is so important that it warrants an illustrative digression before we return to the various studies that have constructively utilized Richardson's mistakes. A major reason for the pitfall of irrelevance is the effort of political scientists to enrich their theory—poor discipline by borrowing the concepts and relationships of physics, biology, and economics. While such analogical reasoning often has proved fruitful, it is also very dangerous unless the political scientist has a healthy regard for the questions he is studying.

Limiting ourselves to one example, let us examine Paul Smoker's application of the equations for damped harmonic motion to the analysis of competitive military spending.[22] Smoker's starting point is an equation describing the oscillations of a particle whose motion is retarded or "damped":[23]

$$(14) \quad m \frac{d^2 s}{dt} + r \frac{ds}{dt} + ks = 0$$

The situation can be exemplified by a particle attached

to a massless spring which is placed in a medium of viscous fluid. In the equation, "s" is the displacement or distance from the position of minimum potential energy (the resting position), "k" is the elastic force of the spring, "r" is the force retarding the motion of the particle, "m" is the mass of the particle and "t" is time.[24]

When the damping force is small compared to the elastic force, then[25]

$$(15) \quad s = Ae^{-bt} \sin(vt + P)$$

where "A", or the initial amplitude, and "P", or the "phase constant," are both determined by the initial position and speed of the particle on the end of the spring and where "v" is the velocity of the particle (ds/dt in equation 14) and "b" is related to "r" in equation 14. Finally, it should be noted that the displacement, s, decreases over time until the motion dies out entirely.

To return to our two preliminary critical questions: What is the political meaning of the terms in equation 15, and what is the political significance of this particular relationship?[26] Smoker uses "s" as a measure of defense expenditures. "A" is the initial value of s, and "t" is time. However, "P" seems to be undefined and "v" is not explicitly characterized in a political way. It should be stressed that we are not questioning whether these terms have been operationalized and measured, but whether they mean anything.

If we assume that all of these variables and constants do have referents in the realm of international politics, we must still ask why equation 15 has been chosen rather than any of the other equations which physicists use. As stated, this equation seems to require defense spending to rise, fall, and go through zero. Clearly this does not happen.

Actually this issue is complicated by Smoker's peculiar choice of axes.[27] For y greater than zero, y is an increasing positive function of United States military expenditures. Below the line y = 0, increasingly negative values of y correspond to increasingly positive levels of Soviet spending. Thus expenditures can never become negative or zero.

Alternatively, this issue might have been dealt with, using ordinary axes, by using the square of the sine function or by adding a constant to the equation resulting in sin(vt + P) + k where the constant k is large enough to raise the whole curve above the line s = 0. If one definitely wanted to use a sine curve, one might even have used it in a formulation that characterized the rate of increase rather than the absolute level of spending as an harmonic phenomenon.[28]

However, the essence of the problem is not solvable with mathematical manipulations. The real issue is that an oscillating function has been chosen, but no meaningful political oscillations have been demonstrated. If the data in this field were sufficiently good and numerous, then a good data fit for an equation might cause us to rethink our concepts. But given the quality of our data it is not too difficult to obtain high correlations for any number of functions borrowed from the repertories of other fields. It is, therefore, absolutely essential that a mathematical model begin with a clear verbal formulation of the relevant issues and that it end with equations that are substantively reasonable. In short, in mathematical political analysis as in the more qualitative approaches plausible arguments based on solid thinking are necessary. Only when these requisites are present is it worthwhile to enquire into the mathematical details of the system.

The Richardson Tradition Extended: Caspary's Models

A number of published models do meet these basic criteria. Most of them may be seen as attempts to over-
come Richardson's dilemmas through the introduction
of more realistic assumptions and more complex formulations. As we shall see, each of these studies captures
some aspect of the armament race phenomenon but
leaves other facets untouched. Little more could be expected, since scholars simply do not know which characteristics of this process are fundamental and which may
be neglected. In this light it is understandable that few
of these studies have employed actual data to test and
apply their conclusions except in a very rough way. The
priority has rightly been the achievement of a general
understanding of the relevant processes rather than precise budgetary predictions.

One of the most interesting models in the Richardson
tradition has been developed by William R. Caspary.[29]
Noting Richardson's failure to derive a "rivalry model,"
Caspary reformulates his predecessor's equations in
order to incorporate the role of relative differences in
armed strength as an arms race stimulus. First, equations
1 and 2 are rearranged to yield[30]

(16) $dx/dt = a(ky/a - x + g/a)$

(17) $dy/dt = b(lx/b - y + h/b)$

The terms dx/dt and dy/dt are still the respective rates
of change in the arms levels of states X and Y. However,
the expressions in parentheses may be interpreted as
stating that a nation increases its military strength in
response to the difference between some proportion of
its existing force level and the strength of its opponent.

In this reformulated system, the various constants no
longer have the same *empirical* referents as in the Rich-

ardson model, although they are *mathematically* equivalent. In particular, k/a, l/b, g/a and h/b are to be seen as new parameters rather than as ratios of the Richardson terms. In equation 16, if we let $y = 0$, then $dx/dt = 0 = a$ $(-x + g/a)$ or $x = g/a$. That is, g/a is now the minimal acceptable force level which is desired by nation X even when its opponent, nation Y, is completely disarmed. This could be either because X is aggressive or because X requires g/a for the maintenance of internal order. h/b in equation 17 is similarly defined.

We must also redefine k/a and h/b. The former may be taken to measure the way in which nation X views the arms level of nation Y. If X is interested only in its own defense and if defensive capability does not require as high an arms level as offensive ability, then k/a would be small. If nation X desires to conquer Y then k/a would be large. In short, this term expresses some combination of fear, suspicion, and aggressiveness.

If we rewrite equations 16 and 17 as shown in 18 and 19[31] then one of the consequences of these manipulations becomes clear: the economic constraints on military spending that Richardson built into his system have now disappeared.

(18) $dx/dt = a(ry - x + y_0)$

(19) $dy/dt = b(r'x - y + y_0)$

$$\text{where } r = k/a,\ x_0 = g/a$$
$$r' = l/b,\ y_0 = h/b$$

In the two alternative models which he presents, Caspary reintroduces the cost constraint—but with the additional assumption that military budgets are subject to diminishing returns. This follows the very plausible hypothesis that a nation is less likely to increase its arms expenditures from 85 to 90 percent of its total budget than it is to raise it from 25 to 30 percent.

The first of these models, to which we must limit ourselves for the sake of brevity, is as follows:[32]

(20) $p \, dx/dt = a \, (C - Mx) \, (1 - e^{-ND/C})$

(20a) $D = ry - x + x_0$

(21) $p' \, dy/dt = a' \, (C' - M'y)(1 - e^{-N'D'/C'})$

(21a) $D' = r'x - y + y_0$

Definitions of terms:

C, C′ : total resources that could be spent on military items.

a, a′ : a measure of the "catching up time" as in Richardson.

D, D′ : the desired arms increases, i.e. what one would like to have if cost were not a consideration.

N, N′ : the costs of *new* arms procurement per armament unit.

M, M′ : the cost of maintaining one unit of old armed force.

p, p′ : dimensional constants used to convert from arms/time to dollars/time.

r, r', x_0, y_0 : these are defined above.

Perhaps the first thing to note about this model is that the concept of total arms expenditures in Richardson's model is now disaggregated into the cost of maintaining existing forces and the expense of obtaining new armaments. This is appealing because an arms *race* presumably consists of new military acquisitions. However, one problem with this formulation is that the maintenance of a constant level of military strength requires the replacement of obsolete weapons systems. But in the process of replacement, the quality of the new arms may represent such an advance that a "qualitative arms race" ensues.[33] What is needed, then, is some distinction between qualitative and quantitative changes in weapons stockpiles.

As for the other terms, $C - Mx$ and $C'y$ represent the "amount of resources available for new procurement" or what remains from the total possible military budget after maintenance costs have been deducted. The second terms, $1 - e^{-ND/C}$ and $1 - e^{-N'D'/C'}$, are used because they conveniently express the idea of diminishing returns. Inclusion of these terms causes equations 20 and 21 to reduce to the Richardson system when x and y are small. That is, when the levels of armaments are low, diminishing returns do not apply.[34]

When the exponent ND/C is moderately large, signifying that what one wants to spend on new weapons is a large fraction of one's total military funds, then the actual increase in military spending becomes less than the desired increase.[35] That is, pdx/dt becomes less than d(ND)/dt. As the desired increase in spending becomes very large, pdx/dt approaches C, which is the absolute ceiling.[36] In summary, the Caspary model states the very reasonable proposition that competitive increases in military spending become progressively smaller as they become too expensive in relation to what the participants are willing to allocate to weaponry.

An interesting aspect of this system is that its equilibrium and stability conditions are the same as those of the Richardson equations except for one extra point.[37] A glance at equation 20 shows that equilibrium occurs (i.e. the derivative is equal to zero) where $C = Mx$ and similarly for equation 21. That is, the arms race stops where both states are spending as much as they can. Alternatively, equilibrium occurs where
$$1 - e^{-ND/C} = 0, \quad 1 - e^{-N'D'/C'} = 0$$
Taking the first equation,
$$1 = e^{-ND/C}$$
and using natural logarithms,

$$0 = -ND/C$$
$$\text{or } D = 0$$

This last condition is precisely the same as in the Richardson equilibrium situation. Furthermore, as Caspary shows, the conditions determining the stability of the equilibrium are also the same.

The interpretation of this finding which Caspary offers is based on the fact that only desired increases (and not total spending) affect diminishing returns in this model since the exponent of e is ND/C rather than (ND + Mx)/C. This may be plausible if we agree that

> Once an arms level is reached it is maintained unless a reduction in the opposing force makes part of it militarily superfluous (i.e. unless D is less than zero). Since any spending program has a certain inertia due to vested bureaucratic and clientele interests this may be an accurate picture of reality.[38]

To show that the nature of economic constraints does make a difference in the course of arms races, Caspary proposes an alternative model in which diminishing returns operate both in new procurement and on maintenance costs. Limitations of space prevent a presentation of this system.[39]

What has been and what might be learned from Caspary's analysis? Although these equations have not been operationalized and tested, it seems evident that both could be done. However, to belabor a point already made, even if we could assume that the relevant parameters have remained constant from 1946 until the present and even if all the annually published statistics were reliable there would probably not be sufficient data to establish a clear preference for one of Caspary's two models.

Nevertheless, let us imagine that all of these mathematical formulations were purely verbal. Such questions as whether arms races are reaction processes and whether they exhibit diminishing returns on new or total procurement are extremely important topics for any mode of analysis. We generally would not reject the insights of a nonmathematical study because it lacked accurate fiscal data. If it did nothing more, Caspary's study would be important because it shows how different kinds of cost constraints could result in differing levels of arms accumulations. It would thus be useful to conduct interview and documentary studies in order to assess the relative impact of various economic constraints on military budgets. If, for instance, we found that elites are more concerned with total weapons costs during recessions or times of increasing demands on government resources, this might lead to more complex modeling efforts. The fact that Caspary's model suggests a variety of such questions is a good indication of its utility.

The Cold War Perspective: Wolfson

A major benefit of the kind of analysis undertaken by Richardson is that theoretical inadequacies can be explicitly identified, thereby facilitating corrective measures, for instance Caspary's. In a study by Murray Wolfson,[40] a rather different alteration in the Richardson model is proposed.

Wolfson starts with the idea that an important characteristic of the post-World War II cold war is that *both* sides are motivated by perceptions of communist success in challenging the status quo: both sides react by increasing their efforts. The West does so in order to counter the communists, while the East increases its activity in order to capitalize on its success. By making the underlying motivations of the arms race asymmetrical, Wolfson departs sharply from Richardson and Caspary.

Before examining Wolfson's model and inquiring into the consequences of his innovation, it is necessary to comment on two other distinctive aspects of his work. The first is that Wolfson proposes his system as a description of the entire cold war rather than just the arms race. Although this frame of reference is historically more limited than some other models examined in this review, it is also more general in that it is not limited to expenditures on military forces. While Wolfson says of his independent variable that it "might be thought of as being measured in terms of dollars expended for military and psychological warfare,[41] it is clear what a wide range of political, economic, as well as narrowly strategic activities are encompassed in his formulations. Nevertheless, the concerns of his model are sufficiently similar to those of the other systems reviewed to allow us to treat it as an arms race model in the Richardson tradition.

The second distinguishing point is that Wolfson's system is posed in difference rather than differential equations. While the mathematical consequences of this are not terribly important, the substantive implications are. Involved here is the manner in which we conceptualize political change. If we return to equations 1 and 2 we see that Richardson's differential equations state that changes in arms levels x and y over time (dx/dt and dy/dt) occur instantaneously as the values of $ky - ax$ and $lx - by$ vary. While this may be a sufficiently good first approximation of reality it is clear that arms levels do not change instantaneously—or even very quickly. In addition, the differential formulation implies that political change is continuous. But in thinking about politics, one must remember the sharp changes from increase to decrease and from constancy to explosion that governmental policies sometimes exhibit.

In contrast, Wolfson's difference equations are cast in terms of discrete changes. Thus equations 22 and 23 state that the activities of each side during year t (Y_t^s and Y_t^u) depend upon the actions of both sides during year t − 1 (Y_{t-1}^s and Y_{t-1}^u). This allows an easy identification of the model's concept of time with such real political time units as fiscal years. Granted that many noncontinuous changes are well approximated by differential equations, the closer fit between the difference formulation and political processes facilitates a more thorough intuitive grasp of the model. The Wolfson equations are:[42]

$$(22)\ Y_t^s = a_s[\gamma_s Y_{t-1}^s - \gamma_u Y_{t-1}^u] + \beta_s Y_{t-1}^u + \kappa_s Y_{t-1}^s$$

for side S
or the Soviet side

$$(23)\ Y_t^u = a_u[\gamma_s Y_{t-1}^s - \gamma_u Y_{t-1}^u] + \beta_u Y_{t-1}^s + \kappa_u Y_{t-1}^u$$

for side U
or the U.S. side

Before defining each of the parameters it will clarify matters to take note of the general meaning of the model. Equations 22 and 23 state that both sides' level of cold war activity result from a linear combination of (1) communist *success*, (2) level of one's opponent's *activity*, and (3) level of one's own *activity*. Thus, the second and third terms are roughly similar to the first and second terms of the Richardson equations. The first term embodies the departure from Richardson which has been discussed.

"Y" denotes the level of cold war activity; the subscript indicates the year and the superscript the side. The "reaction coefficient," "a", measures the effect Soviet success has on each side. The term in brackets shows success rather than activity because the Y_{t-1} variables are multiplied by "efficiency coefficients," "γ", whose magnitude depends upon the degree to which expenditures actually produce the desired results.

Wolfson calls the parameter "β" the "paranoia coefficient" since $\beta_s Y_{t-1}^u$ and $\beta_u Y_{t-1}^s$ show the reaction of one side to the activity of the other regardless of the impact of the activity. Finally the "κ" denote inertia: $\kappa_s Y_{t-1}^s$ and $\kappa_u Y_{t-1}^u$ show the tendency to formulate budgets "incrementally" by using the previous year's expenditures as the baseline.

These equations raise a variety of serious questions. The most fundamental involves the concept of "communist success" conceived of as alterations in the status quo in favor of the Soviet side.[43] To take a presently important example, the Vietminh apparently regarded South Vietnam as about to become communist in the mid-1950s while the West saw the new status quo as establishing a noncommunist regime.[44] This illustration would also indicate that the importance of an area in relation to the world status quo is a function of time: in 1970 both sides would regard a communist victory in South Vietnam as a major change while this might not have been so in 1954.

How serious is this definitional problem? At least in the opinion of this reviewer, Wolfson's innovation does capture an important aspect of contemporary international affairs. It seems clear that since the Second

World War the United States has seen its role as the defender of the status quo against communist encroachments, while the Soviets and Chinese have interpreted history as favoring a changed world order. Even if this is only part of the truth and even if we cannot precisely measure the relevant concepts, it is clearly a contribution to determine the qualitative consequences of this view of the world and to check these consequences, however roughly, against events in the real world.

Another difficulty Wolfson recognizes is that a two-bloc model has become decreasingly relevant in recent years—a problem for all of the two-bloc models reviewed in this paper. Nevertheless, Wolfson makes this simplification because it seems valid for part of the post-World War II period and because greater complexities can be introduced if the initial model proves interesting.[45] The final point to be made before examining the implications of the model is that the equations lack a cost constraint. This seems to be because the model is not strictly focused on military spending but encompasses the latter as well as such cheaper activities as propaganda. The absence of such an inhibitory term must, however, be kept in mind when applying the system to competitive arms expenditures.

For purposes of comparison we shall interpret Wolfson's system as an arms race model and, as in the cases of the Richardson and Caspary equations, we shall enquire about the equilibrium and stability conditions of the system. It should be remembered that an equilibrium point is one where the arms levels for successive time periods are constant. "Stability" implies that if the arms levels during some period under consideration either do not begin at the equilibrium value or are displaced from it by influences exogenous to the model, then they converge toward the equilibrium value. Simply put, the equilibrium and stability conditions tell us under what circumstances the arms race will cease—at least according to the model.

In the interests of brevity we shall only allude to Wolfson's demonstration of his stability conditions.[46] Equations 22 and 23 may be stated in matrix notation as

$$(24) \quad \begin{pmatrix} Y_s \\ Y_u \end{pmatrix}_t = \begin{bmatrix} a_s\gamma_s + \kappa_s & \beta_s - a_s\gamma_u \\ a_u\gamma_s + \beta_u & \kappa_u - a_u\gamma_u \end{bmatrix} \begin{pmatrix} Y_s \\ Y_u \end{pmatrix}_{t-1}$$

Let $\gamma_s = \gamma_u = 1$ because not knowing which side is generally more efficient in its use of resources it is safer to assume equal efficiency. The square matrix in equation 20 now reduces to

$$(25) \quad \begin{bmatrix} a_s + \kappa_s & \beta_s - a_s \\ a_u + \beta_u & \kappa_u - a_u \end{bmatrix} \equiv A$$

Let us define the "trace" of A as the sum of the elements along the diagonal from the upper left corner of the matrix to the lower right. It can be demonstrated that the absolute value of the trace of A must be less than two if the system is to be stable.[47]

That is, $|a_s - a_u + \kappa_s + \kappa_u| < 2$

or $-2 < a_s - a_u + \kappa_s + \kappa_u < 2$

hence (26a) $\dfrac{a_s - a_u}{2} < 1 - \dfrac{\kappa_s + \kappa_u}{2}$

and (26b) $-1 - \dfrac{\kappa_s + \kappa_u}{2} < \dfrac{a_s - a_u}{2}$

It can be shown that inequalities 26a and 26b require that $\kappa_s + \kappa_u < 2$.[48] Indeed, Wolfson assumes that these inertia coefficients are each between zero and unity, which is especially plausible if we interpret these parameters as embodying some aspects of a cost constraint.

That is, we may assume that in the absence of any external reason for increasing cold war expenditures, the Soviets and the Americans would respond to domestic needs by not going above the previous year's military budget.

With the inertia coefficients thus constrained, and with all parameters assumed to be positive, inequalities 26a and 26b imply that

$$(27) \quad -2 < \frac{a_s - a_u}{2} < 1$$

Since Wolfson proposes that the reaction coefficients are likely to be quite large,[49] changes of a few percentage points in one or both could mean a large absolute change and, therefore, a shift to instability. In short, this amounts to saying that, in addition to fear, aggressiveness, and the other influences included in earlier models, the standards of evaluation used by both sides can have a great impact on the possibility of ending an arms race. Other things being equal, stability is fostered if both sides agree on what constitutes a change in the status quo and react in very similar manners to their perceptions.

The discussion thus far has avoided some very serious defects in the model.[50] For instance, in examining the other two stability conditions[51] of the system we have presented, Wolfson assumed that β_s and β_u are considerably smaller than a_s and a_u.[52] In certain situations, this is quite plausible. When the Soviet Union installed missiles in Cuba, the importance of the action was related not to the number or cost of the missiles but to the psychological impact of their location. While $\beta_u Y_s$ may have been small, $a_u \gamma_s Y_s$ was very large. However, it can be argued that for the deployment of intercontinental missiles with a first-strike capacity there is no difference whatever between the simple reaction to the opponent's action (βY) and the assessment of the act's consequences ($a\gamma Y$) for the international strategic situation. This is particularly true of the strategic impact of the numbers of intercontinental missiles based respectively in the U.S. and the USSR. Since, then, Wolfson's assumptions concerning a and β break down for important situations, it would be useful to substitute more refined terms and to consider separately different types of weapons.

Despite these problems, Wolfson's distinction between "effort" and "success" as arms race influences is a very illuminating one. This can be seen by considering that there are many areas in the "third world" where local political conditions may allow very small investments of Soviet or Chinese resources to foster significant changes in the status quo. That is, $\beta_u Y^s$ is not tightly related to $a_u \gamma_s Y^s$.

Wolfson's model is valuable in much the same way as is Caspary's. In both cases, the initial equations embody intuitively plausible interpretations of arms race phenomena. The logical implications of these formalized intuitions are then deduced. Some deductions shed light on competitive military spending, while others indicate defects in the model. The latter are often sufficiently explicit and interesting to stimulate further analysis.

An Application of Data: Milstein and Mitchell

An exception to the tendency not to fit equations to data is represented by Jeffrey S. Milstein and William C. Mitchell's study of the pre-World War I naval arms race.[53] Milstein and Mitchell utilize a suggestion by

Richard Lagerstrom and Robert North[54] that, in determining how much to increase their military expenditures, government leaders estimate the future strength of their opponents and adjust their funding accordingly. But a calculation of a nation's future position requires a consideration not only of how much armed strength it has but also how fast that armament level is growing. The final response to these estimates would be an adjustment in the rate of increase of one's military strength—that is, an acceleration or deceleration in the rate of procurement.

These ideas are incorporated in a system of second-order differential equations:

(28) $\dfrac{d^2 x}{dt^2} = ky + m\dfrac{dy}{dt} - ax - c\dfrac{dx}{dt} + g$

(29) $\dfrac{d^2 y}{dt^2} = lx + n\dfrac{dx}{dt} - by - d\dfrac{dy}{dt} + h$

The meaning of these equations is as follows: "$d^2 x/dt^2$" and "$d^2 y/dt^2$" indicate the rate of acceleration in arms spending. That is, they define the "rate of increase of the rate of increase" in military levels. "k" and "l" have the same definitions as in the Richardson equations: they are the "reaction coefficients," which show the degree to which each side reacts to the other's absolute level of forces. "m" and "n" also measure reaction, but to the opponent's rate of increase of weapons expenditures.

Finally, "a", "b", "c", and "d" are cost or fatigue coefficients. The first two cost terms show the burdensomeness of maintaining one's absolute level of forces, the second pair indicate the reluctance to increase expenditures. Lastly, "g" and "h" are the same measures of grievance and hostility as the corresponding terms in the Richardson system.

As in our previous critiques, an important question is how relevant to arms race processes are the basic assumptions of this system. The idea of including second-order derivatives so as to express the effects of budgetary increases as well as budgetary levels is clearly a good one. Indeed, it seems likely that in many cases a government will react to its enemy's military buildups but will not be concerned about a constant force level that has acquired legitimacy through time. Similarly, a legislature or a public may be willing to continue funding its military at the customary level but might oppose increases except in times of unusual tension.

We must also ask whether the new assumptions carry new implications concerning the cessation or escalation of arms races. At the point of equilibrium, both sides' levels of military spending are constant. Since both the first and the second derivatives of a constant are zero, equations 28 and 29 become:

(30) $0 = ky - ax + g$

(31) $0 = lx - by + h$

Calling these constant x and y "x_0" and "y_0" we have

(32) $x_0 = \dfrac{hk + gb}{ab - kl}$

(33) $y_0 = \dfrac{gl + ha}{ab - kl}$

These are precisely the equilibrium conditions of the Richardson model as discussed earlier.

That the difference between the Richardson equations and the Milstein-Mitchell model does not affect the existence and stability of the equilibrium point may seem puzzling. However, *all* of the parameters influence the *time* it takes to *reach* equilibrium.[55] This impact is extremely important, as we shall see after examining the empirical application of this model.

In testing their model, Milstein and Mitchell employ naval budgets from 1865 to 1914. Their technique is to use the empirical values of their independent variables over, say, five points in time in order to predict the dependent variable. Stepwise regression is used for this purpose. Next, the value of each of the independent variables for the sixth period is predicted from the five real data points. The five values for time t_2 through t_6 (i.e. four real and one predicted value) of each of the independent variables are again used to predict the independent variable. After the first five time periods, only artificial data are being used. The reason for this procedure is to obtain a more stringent test of the model than would be provided by such measures of association as correlations. As can be seen from figure A.4, the fit between the actual and the predicted British naval expenditures is extremely good.

One likely reason why Milstein and Mitchell's system so successfully describes the pre-World War I arms race is that the international situation during that period was more stable for a longer time than has been true since 1918. As has been remarked earlier, it is probable that the influences relevant to arms races have changed radically and frequently since World War I.

Equations 34 and 35 show the Milstein-Mitchell regression estimates of the parameters appearing in equations 28 and 29.[56] It should be noted that in 34 and 35 a single (′) indicates the first derivative while (″) denotes the second derivative.

(34) $B'' = 0.10 G + 0.33G' - 0.08B - 0.58B'$ for Britain

(35) $G'' = 0.32B + 0.26B' - 0.33G + 0.05G'$

for Germany and Britain's other rivals

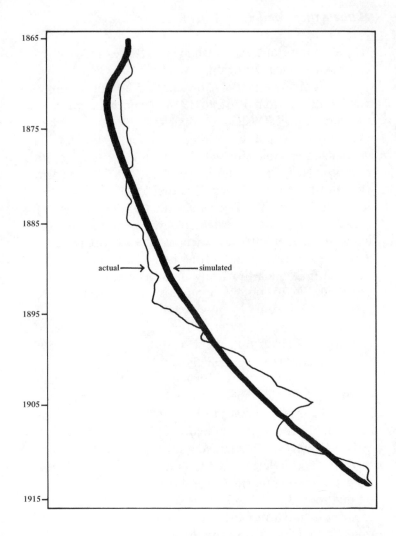

Figure A.4
Source: Milstein and Mitchell, "Computer Simulations of International Processes," p. 123.

The authors offer several interesting interpretations of these results. One is suggested by the fact that in equation 34 the coefficient of G' is greater than that of G, while in equation 35 the coefficient of B is greater than that of B'. That is, Germany reacted more strongly to the absolute naval strength of Britain, while Britain was more influenced by how fast Germany was catching up (by how large G' was). The authors aptly compare this difference in attitudes between leader and follower to the 1968 presidential elections:

> That Hubert Humphrey was closing fast in the last week was of little value to him—his concern was that he was behind. Richard Nixon, on the other hand, reacted sharply to Humphrey's closing rush.[57]

Finally, it is interesting to ask whether this model, with the parameter values shown in equations 34 and 35, implies that the arms race would have stopped if World War I had not intervened. As can be shown with the aid of Routh's criterion, equations 34 and 35 do state that eventually this competition would have ceased.[58] This finding, combined with the fact that the war did break out, should convey an important lesson: formal models of politics must be applied to reality with care.

As a general matter this last problem can result from two circumstances. If the time required to reach equilibrium is very long, then the world situation may change drastically in the interim, thus making the model irrelevant. In addition, if the equilibrium is at a level of armaments so high that unbearable international tension results, then war may well occur before the arms race stops. In brief, the stability implications of a model indicate the conditions under which an end to an arms race is possible—at least according to the model. An estimation of whether and when such a halt would actually take place must depend upon precise knowledge of the situation and upon assumptions concerning the changes that are likely to occur.

Strategic-Probabilistic Models

The models coming under the general rubric of the "Richardson tradition" vary widely in their assumptions, their implications, and their quality, as we have tried to show. They do, however, share certain basic characteristics. In all cases they are "deterministic" in the sense that the concept of probability is not explicitly included. Decision-makers are pictured as reacting automatically to such variables as their own levels of military spending and those of their opponents. Also missing from the Richardson tradition is an explicit formulation of the role of strategy. Strategy is, however, implied by the fact that the reaction and grievance terms would be larger for an aggressive nation than for one interested only in deterrence.

How important are these two concepts in the analysis of competitive military spending? As Martin McGuire and others have shown,[59] governments pay not only for the ambitiousness of their strategic goals but also for some acceptable, minimum degree of confidence of attaining them. To take a simple hypothetical example, a belligerent policy might require more missiles, and thus be more costly, than a deterrent policy. In the same vein, the greater the number of missiles one wants to survive a first strike, the more one must spend on some combination of defenses for and quantities of missiles. But, in addition, if a government required a 90 percent probability of success, it would have to spend more on missile numbers and defensive facilities than if a 60 percent probability would suffice.

These kinds of issues receive prominent attention in the analyses of Arthur Lee Burns, McGuire, and others.[60] In the interest of brevity, we shall illustrate this approach with only one short example drawn from McGuire's excellent study of the contemporary arms race, namely, missile buildups. Unfortunately, the depth and variety of McGuire's analysis prohibits anything like a comprehensive review in this essay.

Governments spend money on missiles in order to deter attack, or to create an aggressive potential, or for both reasons. One can, then, start with the various strategic stimuli to military stockpiling and then introduce technological, economic, and political considerations into the model. In this manner, McGuire makes certain simplifying assumptions in order to specify the probability that a given side's strategic goals would be fulfilled in the event of an attack.[61] If a side were only interested in deterrence, the objective would be to have at least some specified number of missiles survive an enemy's first strike. The desire for an aggressive capability would require the ability to destroy some proportion of the opponent's retaliatory force. Or both motives might be present.

From his probability equations, McGuire deduces an arms race model which is "interactive" in the same sense as are the models we have already examined. That is, each side is pictured as planning its missile procurement in the light not only of its own goals and capabilities but also those of its opponent.[62]

$$(36) \quad z_y = t_a [ys_y{}^{x/y}(1 - s_y{}^{x/y})]^{\frac{1}{2}} + ys_y{}^{x/y}$$

(37) $z_x = t'_a [xs_x^{y/x}(1 - s_x^{y/x})]^{1/2} + xs_x^{y/x}$

where the terms in equation 36 are defined as follows: (Corresponding terms in equation 37 are similarly defined)

v_y :   the number of Y's missiles surviving attack.

z_y :   the minimum number of Y's missiles which Y wants to survive (if equation 36 describes Y's decision process) or the maximum number of Y's missiles which X wants to survive (if equation 36 describes X's decisions).

t :   the point on the abscissa of the normal curve which determines the probability that $v_y \leqslant z_y$.

s :   the probability that a single Y missile site will survive an attack by one of X's missiles.

x :   the number of X's missiles.

y :   the number of Y's missiles.

a :   the proportion of x actually fired.

If equation 36 represents the decisions of side Y and 37 those of X, then we have a situation in which both sides are interested only in deterrence. They are therefore concerned with maximizing their own surviving missiles in case of a first-strike, counterforce attack. Corresponding to these assumptions, z_x is labeled R_x indicating that this is X's "reaction curve": i.e. for any value of y, X procures a number of missiles that is sufficiently large to attain a specified probability that z_x will survive an attack. R_y is similarly defined.

As can be seen from figure A.5, there is a stable equilibrium at x_0, y_0. If Y deploys y_2 missiles, X can deploy some smaller number, between x_2 and x_0, and still reach its reaction curve. At X's new strength, Y is now spending too much and therefore reduces its missile force. The process continues until equilibrium is

reached. Similarly, there is a tendency to increase missile strength from levels below the equilibrium. This
situation fulfills the intuitive expectation that if two
opposing blocs distrust each other and want only to
defend themselves, then some condition between complete disarmament and explosive escalation is likely.
However, the shape of the curves and McGuire's assumptions concerning military technology show that this conclusion depends upon the premise that defense has an
advantage over offense. Conditions favoring a first strike

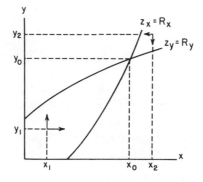

Figure A.5
Source: McGuire, *Secrecy and the Arms Race,* p. 145. Copyright 1965 by
the President and Fellows of Harvard College.

could lead to a much more unstable situation even
though both sides only want to act defensively.[63]

Using this mode of analysis, McGuire also shows that
if both sides want a first-strike ability, then there is an
equilibrium point, but it is unstable—any displacement
from it leads to unlimited escalation. In addition, the
more realistic situation in which each side desires both
deterrent and first-strike abilities is explored. Finally, in
a lengthy examination of the role of secrecy, McGuire
explores the effects of different levels of information on

military expenditures and missile stockpiling.[64]

McGuire's study is essentially a qualitative analysis of one aspect of arms races. The result is a clearer understanding of the way in which a variety of influences can foster such competitions. As is also true of Burns and Intriligator, whose approaches are somewhat similar, this contribution is made without the aid of real data. Here again the intent is to clarify and organize concepts and relationships and to determine their implications.

An Overview

Several observations on the mathematical arms race literature can be offered. First, it seems clear that no extant models are capable of accurately predicting levels or growth rates of military spending in the contemporary world. This is due to two types of deficiencies: poor data and inadequate theory. Secondly, several important contributions have been made to our qualitative understanding of arms races. Finally, the latter kind of insight can and should be used to suggest exploitable, but largely untapped, sources of data that might help remedy the quantitative problems described in this paper.

Of the empirical difficulties, the quality of the data is the most obvious; another is the quantity of data from any period for which the particular model is relevant. In examining the work of Milstein and Mitchell, I suggested that they were warranted in using nearly fifty annual observations because they could reasonably assume that influences depicted in their model had remained constant throughout the pre-World War I period which they seek to explain. On the other hand, I contended that

Richardson was not justified in using post-1918 data to infer a pre-1914 value for his reaction coefficient "k" since the world had changed very drastically in the interim. However, not only must such parameters remain constant throughout the time span of the analysis, but the particular relationships imbedded in the structure of the equations must be valid for the entire period from which the data are drawn. For example, the form of Wolfson's system is based on the assumption that both sides take Soviet success as their standard. Clearly the model as a whole, and not merely its parametric values, would have to be altered in order to extend it to a time in which each side looked to the initiatives of the other.

This argument implies that the twenty-odd fiscal years since the end of the Second World War offer the *maximum* number of observations presently available to test models of twentieth-century arms races. This situation is bad enough in view of the complexity of some of the models reviewed in this paper. But one has only to consider such major changes as the end of tight bipolarity and the rise to international importance of guerrilla wars in the "third world" to realize that the even greater complexity required of a mathematical analysis of the contemporary period would necessitate a huge amount of data. However, the picture should not be overdrawn. It takes neither a large amount of data nor complicated techniques to reject a model whose concepts make no sense or whose predictions are grossly unrealistic. But this leaves a very large number of possibly useful systems that are statistically indistinguishable. Nevertheless, as we have argued, there are other grounds upon which preferences could be based, such as relevance and plausibility.

Before leaving these empirical issues, it should be

noted that a very important source of data has gone
unmentioned in this review: domestic politics. Formal
studies of arms races have left almost completely untouched such matters as domestic interest groups, public
opinion, etc. This is true despite the appearance of such
vague terms as "cost constraints" and "grievance coefficients." Obviously these intranational influences are
vitally important in determining the size of defense
expenditures, and their inclusion in formal models
would do much to increase our understanding of arms
race processes. In addition, since statistics on national
politics are not limited to annual appearances, the inclusion of these phenomena would greatly enhance our
ability to test analyses of competitive military spending.
True, hard data on domestic influences in communist
countries are almost as difficult to obtain as strategic
information. Nevertheless, the further illumination of
buildups in the armed forces of the Western powers
would be no mean contribution. Even in the case of
Soviet expenditures, a model that offered a roughly
plausible explanation of a number of influences might
engender more confidence than one that quite reasonably encompassed only a few variables.

In addition to these fairly obvious problems with data,
another hindrance to accurate predictions of armament
expenditures is the presently inadequate condition of
relevant theory. It is not only the case that the important variables cannot now be measured—we do not
really know which variables are important! For reasons
that should now be evident, we cannot rely solely upon
existing data to sort this matter out. However, the kind
of qualitative analysis that has been reviewed can help
to remedy the situation. For example, Caspary's study
indicates that a distinction with important consequences

is whether policy-makers are inhibited from further increasing their weapons expenditures by what they are spending on new procurement or by their outlays both for new forces and for the maintenance of existing armed forces. Interviews with congressmen and Defense Department officials, studies of congressional voting records, analyses of Soviet speeches and economic plans all might help in deciding this and similar questions.

Finally, one of my major contentions has been that even if empirical problems made it permanently impossible to achieve accurate predictions in this area, this kind of analysis would be of great value. A major contribution these models can and should make is the organization and clarification of commonly used verbal theories and the illumination of these formulations' often unsuspected implications. In essence, this amounts to a more precise determination of what we mean when we think of arms races in particular ways. An example of such clarification is McGuire's demonstration of the effects of different technological conditions, various strategic preferences, and differing levels of information. Similarly, through the qualitative analysis of stability and equilibrium conditions, several of the studies reviewed here indicate a variety of circumstances that could cause explosively escalating arms races. Even without hard data, an attempt to synthesize the disparate assumptions of these analyses might lead to a greater understanding of competitive military spending. Simulation is among the techniques that might be tried.[65]

NOTES

Chapter 1

1. Institute for Strategic Studies, *The Military Balance, 1968-1969* (London: Institute for Strategic Studies, 1968).

2. Charles L. Taylor and Michael C. Hudson, *World Handbook of Political and Social Indicators* (2d ed. New Haven: Yale University Press, forthcoming.

3. *The Federal Budget: Its Impact on the Economy,* Fiscal 1970 edition (1969). Cited in Frank A. Pinner, "The Utility of Utility: Policy, Decision-Makers, and Individual Choice," paper presented to the annual meeting of the American Political Science Association, New York, 1969.

4. The Civil War figures include only Union soldiers, though the population figures include the Confederacy. Data on Confederate military strength are very poor; however, the best available estimates indicate a consistent inferiority of 1 to 2 for Southern forces, a ratio which grew more adverse after mid-1863. Use of such a ratio would put Union plus Confederate military personnel at no higher than 17 percent of the male labor force, still far below the World War II mark.

5. Malcolm W. Hoag, an economist not unsympathetic to defense needs, says that after the early sharp cuts of the Eisenhower administration military outlays remained "surprisingly high" because the economies implicit in the new look doctrine were never realized. See "What New Look in Defense?" *World Politics* 22, no. 1 (October 1969): 2.

6. *Public Expenditures in Communist and Capitalist Nations.* (Homewood, Ill.: Irwin, 1969), p. 113.

7. See Lewis F. Richardson, *Arms and Insecurity* (Pittsburgh and Chicago: Boxwood and Quadrangle, 1960), pp. 14-16.

8. The problem is especially acute with those models derived from physics, for example, Paul Smoker, "Trade, Defense and the Richardson Theory of Arms Races: A Seven Nation Study," *Journal of Peace Research,* 1965, no. 2: 161-76, and "The Arms Race: A Wave Model," *Peace Research Society (International) Papers* 4 (1966): 151-92.

9. *Public Expenditures in Communist and Capitalist Nations,* pp. 111-14.

10. Taylor and Hudson, *World Handbook.*

11. "Budget Alternatives After Vietnam," in Kermit Gordon, ed., *Agenda for the Nation* (Washington: Brookings Institution, 1968), pp. 41-42.

12. See also the related argument about excess capacity by Alfred Kahn, "The Tyranny of Small Decisions: Market Failures, Imperfections, and the Limits of Economics" in Bruce M. Russett, ed., *Economic Theories of International Politics* (Chicago: Markham, 1968).

13. A recent and sophisticated example is Harry Magdoff, *The Age of Imperialism: The Economics of U.S. Foreign Policy* (New York: Monthly Review Press, 1969).

14. This position is adhered to by most economists. See, for example, Emile Benoit and Kenneth E. Boulding, eds., *Disarmament and the Economy* (New York: Harper, and Row, 1963); U.S. Arms Control and Disarmament Agency, *Economic Impacts of Disarmament* (Washington: U.S. Gov-

ernment Printing Office, 1965); Emile Benoit, "The Monetary and Real Costs of National Defense," *American Economic Review* 58, no. 2 (May 1968): 398-416, and Walter Isard and Eugene Schooler, "An Economic Analysis of Local and Regional Impacts of Reduction of Military Expenditures," *Peace Research Society (International) Papers* 1 (1964). M. Marantz, "Dépenses d'armement et économie nationale," *Revue française de Sociologie* 2, no. 2 (June 1961): 54-65, also concludes that military expenditures do not constitute a necessary element in American prosperity. He has a table of changes in American defense and in GNP over the years 1924-60, where one can see that of the eight years when defense fell as a proportion of GNP, in only four did total GNP go down. The classic statement on 19th-century imperialism is William L. Langer, *The Diplomacy of Imperialism* (New York: Knopf, 1956).

15. Eugene Staley, *War and the Private Investor* (Garden City: Doubleday, 1935), nevertheless maintains that more often the causal chain runs the other way: "International friction over private investments has been a good deal more frequent and dangerous where private investments have been pressed into service as instruments, tools, of a larger political purpose which the investments themselves did not originate. Investments used in the quest for national glory, and the like, have been more productive of international friction in the past than investments actuated solely by private profit motives." (pp. xv-xvi)

16. *Congressional Record,* June 2, 1969, p. 114384.

17. *The Military-Industrial Complex: A Problem for the Secretary of Defense* (Washington: *Congressional Quarterly,* May 24, 1968), p. 1178. See also Carl Kaysen, "Military Strategy, Military Forces, and Arms Control," in Gordon, ed., *Agenda for the Nation,* pp. 549-84. For some other estimates of possible reductions and discussion of why their achievement may be politically difficult, see Jonathan B. Bingham, "Can Military Spending Be Controlled?" *Foreign Affairs* 48, no. 1 (October 1969): 51-66.

18. *A New Administration Faces National Security Issues: Constraints and Budgetary Options,* P-3959 (Santa Monica, Calif.: Rand Corporation, November 1968), p. 26.

19. Merton J. Peck and Frederic M. Scherer, *The Weapons Acquisition Process: An Economic Analysis* (Boston: Harvard University School of Business Administration, 1962), p. 129.

20. Ibid., p. 128.

21. Ibid., p. 127. Also, employment in defense industries is more variable than in industry as a whole. See *The Variability of Employment in Defense-Related Industries,* RMC Report UR-021, Prepared for the Office of the Assistant Secretary of Defense (Systems Analysis) Economics (Bethesda, Maryland: Resource Management Corporation, September 1967).

22. Adam Yarmolinsky, "The Problem of Momentum," in Abram Chayes and Jerome B. Wiesner, eds., *ABM* (New York: New American Library, 1969). Peter L. Mullins, "Have We Overreacted to the Influence of the Military-Industrial Complex?" (Columbus, Ohio: Mershon Center for Education in National Security, 1969, mimeo.), however, reports *some*

evidence that the top DoD contractors are diversifying and becoming less dependent on defense orders.

23. U.S. Senate, Committee on Foreign Relations, Subcommittee on Disarmament, "The Economic Impact of Arms Control Agreements," *Congressional Record,* Oct. 5, 1962, pp. 2139-94, and U.S. Senate, Committee on Labor and Public Welfare, report of Subcommittee on Employment and Manpower, *Convertibility of Space and Defense Resources to Civilian Needs: A Search for New Employment Potentials,* 88th Congress 2d Sess. (Washington: U.S. Government Printing Office, 1964). Both are discussed in Marc Pilisuk and Thomas Hayden, "Is There a Military-Industrial Complex Which Prevents Peace?" *Journal of Social Issues* 21, no. 3: 67-117.

24. *The Weapons Acquisition Process.*

25. Reported in Yarmolinsky, "The Problem of Momentum," p. 148. Richard Barnet, *The Economy of Death* (New York: Atheneum, 1969), chap. 2, is also enlightening.

26. See Congressional Quarterly, *Legislators and the Lobbyists,* chap. 4, "The Military Lobby" (Washington: Congressional Quarterly, 1965) for this and similar tales. See also Julius Duscha, *Arms, Money, and Politics* (New York: Ives Washburn, 1965).

27. Vincent Davis, *The Admirals Lobby* (Chapel Hill: University of North Carolina Press, 1967), p. 8.

28. Ibid., p. 258.

29. "How the Pentagon Works," *Atlantic Monthly,* March 1967, p. 61.

30. *An Inquiry into the Causes of the Wealth of Nations* (New York: Modern Library, 1937), p. 658. First published 1776. In "The Political Order and the Burden of External Relations," *World Politics* 19, no. 3 (April 1967): 443-64, Paul Hammond notes that Frederick William II refused to meet Napoleon's *levée en masse* with one of his own for fear of undermining the stability of his own regime; even as late as World War I states' strategy of rapid mobilization and preemption was forced upon them by their inability to support large standing armies.

31. All this public opinion material is from an A.I.P.O. press release of August 14, 1969.

Chapter 2

1. According to Samuel P. Huntington, *The Common Defense* (New York: Columbia University Press, 1961), p. 134, between 1950 and 1958 Congress reduced military appropriations by roughly 3 percent and foreign aid requests by about 18 percent.

2. Vincent Davis, *The Admirals Lobby* (Chapel Hill: University of North Carolina Press, 1967), chap. 1.

3. Elias Huzar, *The Purse and the Sword: Control of the Army by Congress Through Military Appropriations* (Ithaca, N.Y.: Cornell University Press, 1950).

4. See, for example, Congressional Quarterly, *The Military-Industrial Complex: A Problem for the Secretary of Defense* (Washington: Congressional Quarterly, May 24, 1968).

5. Gordon Tullock, "Problems of Majority Voting, *Journal of Political*

Economy 57 (1959): 571-79; also Julius Margolis, "Metropolitan Finance Problem: Territories, Functions, and Growth," in *Public Finances: Needs, Sources, and Utilization: A Conference of the Universities-National Bureau of Economic Research* (Princeton: Princeton University Press, 1961). On the other hand, Anthony Downs maintains that individuals' ignorance of the full benefits they receive from public expenditures helps to keep the government budget smaller than it would be with better information. See his "Why the Government Budget is Too Small in a Democracy," *World Politics* 12, no. 4 (1960): 547-61.

6. There are three partial exceptions in the literature open to the public (good but secret data exist at the Pentagon). One is the set of studies prepared by Walter Isard and his associates: Isard and James Ganschow, *Awards of Prime Military Contracts by County, State and Metropolitan Area of the United States, Fiscal Year 1960* (Philadelphia: Regional Science Research Institute, 1962); Isard and Gerald J. Karaska, *Unclassified Defense and Space Contracts: Awards by County, State, and Metropolitan Area of the United States, Fiscal Year 1962* (Philadelphia: World Friends Research Center, 1964); and Isard and Karaska, *Unclassified Defense Contracts: Awards by County, State, and Metropolitan Area of the United States, Fiscal Year 1964* (Philadelphia: World Friends Research Center, 1966). Others are U.S. Department of Commerce, Bureau of the Census, *Current Industrial Reports: Shipments of Defense-Oriented Industries, 1965* (Washington: U.S. Government Printing Office, July 1967; also other years), and the list of major defense installations (bases and manufacturers) by city or congressional district in Congressional Quarterly, *The Military Industrial Complex.* All of these, however, have serious disadvantages for use in this analysis. Unclassified contracts account for less than 50 percent of all DoD contracts. The prime contract data ignore subcontracting patterns, which are very important in defense industries, as well as contractors' purchases of raw materials or partially finished products. This is serious enough when working at the aggregated state-by-state level but crippling with smaller units. Isard and Ganschow (p. 1) quote the Department of Defense: "Because of the extent to which subcontracting occurs and because precise knowledge is lacking concerning the geographic distribution of these subcontracts, any breakdown of prime contract awards below the State level must be considered to contain a built-in error so great as to obviate the validity of any conclusions." The Isard materials include only prime contracts, and the Department of Commerce figures include only a small fraction of subcontracts, "because the usual subcontracting industries were excluded from the scope of the survey" (p. 2). Both ignore wages and salaries paid directly to military and civilian personnel—a very large item. It is not clear how many subcontractors are represented in the Congressional Quarterly material, but the bulk of them simply are not traceable with existing public information. It would be attractive to use the data because of the congressional district breakdown presented. Nevertheless, defense installations are not given according to dollar value of defense work, only in broad categories of total employment. More importantly, nearly all the major cities have their congressional districts aggregated in these data; hence one must assume, for example, that the impact of the

Brooklyn Navy Yard is as important to a representative from the Bronx as to the congressman in whose district it lies. Given the great complexity and expense of processing House of Representative roll-call data it seemed not worth doing so with data containing so much error. Later studies may, however, find the effort valuable.

7. The basic procedure is named for its developer, Louis Guttman, whose first published discussion was in "The Cornell Technique for Scale and Intensity Analysis," *Educational and Psychological Measurement* 7 (Summer 1947): 247-80. It was first applied to legislative roll-call analysis by Duncan MacRae, *Dimensions of Congressional Voting* (Berkeley and Los Angeles: University of California Press, 1958), and George Belknap, "A Method for Analyzing Legislative Behavior," *Midwest Journal of Political Science* 2 (November 1958): 377-402. See also Bruce M. Russett, "International Communication and Legislative Behavior: The Senate and the House of Commons," *Journal of Conflict Resolution* 6, no. 4 (December 1962): 291-307. A good introduction and bibliography on legislative voting analysis and scaling is Lee F. Anderson, Meredith W. Watts, Jr., and Allen R. Wilcox, *Legislative Roll Call Analysis* (Evanston, Ill.: Northwestern University Press, 1966). In this application, Yule's Q is used as the measure of similarity between roll calls. Its use was developed by MacRae in "A· Method for Identifying Issues and Factions from Legislative Votes," *American Political Science Review* 59 (December 1965): 909-26 and in *Parliament, Parties, and Society in France, 1946-1958* (New York: St. Martin's Press, 1967).

The analysis reported here was performed on the Yale University 7090-94 computer, on SCALE and SCORE programs written by John L. McCarthy of the History Department at Yale. Earlier versions of these programs were written by Herbert Weisberg at the University of Michigan, with an earlier version of the SCALE program in turn written by James Lingoes of Michigan. They are described in McCarthy's "'Voting Alignments and Reconstruction in the House of Representatives, 1863-1875'" (Ph.D diss., 1970). I am very greatly indebted to McCarthy for modifying his program to fit my data, performing the computer runs, and helping me to interpret them. I do not want to burden him with responsibility for the results, but without his work I would not have them.

An alternative procedure for processing large bodies of roll-call data is factor analysis, which I have employed elsewhere, inter alia in *International Regions and the International System* (Chicago: Rand McNally, 1967), chaps. 4 and 5. For this analysis I nevertheless preferred scaling procedures, partly for their relative simplicity and partly because they avoid certain assumptions required by factor analysis which may not always be appropriate. Perhaps most important, the scaling approach avoids the necessity of assigning interval measures to the voting categories (pro, con, and absent). Also, it is always possible but sometimes difficult, with factor analysis, to permit the scales that emerge to be moderately correlated with each other if that is the most "accurate" representation of the underlying voting patterns. This scaling procedure does not make the requirement of independence found in factor analysis with orthogonal rotation.

8. But only slight. The two senators who most often stood alone on

defense issues were Gruening and Morse, who even by the votes employed in the scale stand at the extreme "dovish" end.

9. "Organization of American National Security Opinions," *Public Opinion Quarterly* 33, no. 2 (Summer 1969): 227-29, 235. Italics in original.

10. A rank-order correlation may vary between +1.00, indicating perfect agreement in the ordering of two lists, and -1.00, indicating perfect reversal of rankings. A correlation of 0.0 means that the two listings are completely unrelated. When using a product-moment correlation on interval data one can interpret the strength of the correlation in terms of the proportion of the variance that is accounted for, but unfortunately that interpretation is not possible with a rank-order correlation. Nevertheless some idea of how strong the relationship is with 100 observations can be obtained as follows: A product-moment correlation accounting for 12 percent of the variance is statistically significant at the .001 level; by comparison, a rank-order correlation of .22 or higher is significant at the .001 level. By an .001 level we mean that if two scales were completely unrelated to each other in a population, the odds are only one in a thousand that they would appear in our sample to have a correlation as high as the threshold figure given. Hence we are protected from treating as interesting a relationship that may appear only by chance.

Of course, the "classical" conditions of sample and population are not met with our data; we would be hard put to say what the population of observations is, and we do not have a random sample. Nevertheless, significance tests are useful in gauging the strength of a relationship against the amount of random variation we might expect, and a sophisticated rationale for their use in such situations exists. For a good example, see J. Johnston, *Econometric Methods* (New York: McGraw-Hill, 1963), chap. 1.

11. Franklin A. Long, "The Industrial Impact of Apollo," *Bulletin of the Atomic Scientists* 25, no. 7. (September 1969): 71.

12. Possibly these positions are not quite so inconsistent as they seem, since an important part of the gun control effort has been directed against the *importation* of foreign-made firearms, and the recent federal legislation has resulted in a boon for some domestic small-arms makers.

13. This assertion is of course based on a single Congress only; it might have to be modified on examination of voting during the Eisenhower years, for example. Nevertheless note H. Bradford Westerfield's observation of "a gradual routinization of national security policy from about 1955 to 1965 that markedly reduced the Administration's need to enlist congressional participation in major new departures." In "Congress and Closed Politics in National Security Affairs," *Orbis* 10, no. 3 (Fall 1966): 748.

14. This continuity apparently extends into the 91st Congress. Eighty-four senators carried over from either the 87th Congress or the 90th into the 91st. Dividing the previous general defense rankings approximately in half, only 10 senators' votes on the ABM (Smith Amendment, defeated by a 50-50 tie) would be incorrectly predicted.

Chapter 3

1. This is the primary difficulty with the only previous studies relating defense spending to legislative behavior, the otherwise interesting articles by Charles H. Gray and Glenn W. Gregory, "Military Spending and Senate Voting: A Correlational Study," *Journal of Peace Research,* 1969, no. 1, pp. 44-54; and Stephen A. Cobb, "Defense Spending and Foreign Policy in the House of Representatives," *Journal of Conflict Resolution* 13, no. 3 (September 1969): 358-69. Gray and Gregory used only the aggregate of total defense spending as their attempted explanatory variable, and found no relation with appropriations or authorization votes overall—though they did find some rather modest relationships between military spending patterns and votes on nondefense issues. Cobb found no relation between spending and foreign policy votes. A methodological discussion of why I believe no relationship emerged can be found in my "Defense Spending and Foreign Policy," *Journal of Conflict Resolution* 14, no. 2 (June 1970).

2. U.S. Bureau of the Census, *Statistical Abstract of the United States,* annual issues (Washington: U.S. Government Printing Office). These are not notably different from the figures compiled by Isard and his colleagues referred to in chapter 2, except that the earliest Isard data omitted classified contracts.

3. Charles M. Tiebout, "The Regional Impact of Defense Expenditures: Its Measurement and Problems of Adjustment," in Roger E. Bolton, ed., *Defense and Disarmament* (Englewood Cliffs, N.J.: Prentice-Hall, 1966), p. 130.

4. Bolton, *Defense Purchases and Regional Growth* (Washington: Brookings Institution, 1966), pp. 164-72; *Report of the Independent Study Board on the Regional Effects of Government Procurement and Related Policies* (Washington: U.S. Department of Commerce, 1966), p. 22; and Steger, "Further Thoughts on the Use of Federal Procurement for Preferred Regional Impact," Consad Research Corporation, paper for the Regional Science Association Annual Meeting, Cambridge, Mass., November 1968. The Steger data are for an aggregate of "Priority One" spending by DoD for procurement of goods and services, excluding expenditures for military installations and civil defense stockpiling.

5. Available for June 1967 and June 1968, in an unpublished report from Office of the Assistant Secretary of Defense (Systems Analysis: Economic & Resource Analysis), *Defense-Generated Employment, June 1968* (Washington: no date). This represents an updating of the data contained in Roger F. Riefler and Paul B. Downing, "Regional Effect of Defense Effort on Employment," *Monthly Labor Review,* July 1968, pp. 1-8. The index almost certainly understates the importance of contracts relative to payrolls, perhaps by half.

6. National Aeronautics and Space Administration, *Annual Procurement Report, Fiscal Year 1961* (Washington: U.S. Government Printing Office, 1961; other years as appropriate), and "The Spatial Impacts of Defense-Space Procurement: An Analysis of Subcontracting Patterns in

the United States, *Peace Research Society (International) Papers* 8 (1967): 109-22.

7. Just to be certain, I did in fact use a conglomerate index in the correlations with Senate voting, but as all the correlations were equal to or (usually) less than those for single components, I have not bothered to report them.

8. Again, I did check, but again the political correlations were very similar to but not higher than the *Statistical Abstract* ones, so only the latter are discussed below.

9. With a two-tailed test; that is, we do not dismiss the possibility of negative correlations.

10. It does not help to disaggregate spending by single years; neither for these relationships nor for those in table 3.5 are the relationships ever notably stronger when 1961 and 1962 (or 1967 and 1968) spending totals are used separately than when combined as above, so to simplify the presentation they are not shown.

11. By comparison, Gray and Gregory, "Military Spending and Senate Voting," report as positive findings correlations no higher than −.14 for defense spending with two scales, measuring "liberalism" in general and liberal attitudes on a foreign aid-test ban scale for the 88th Congress.

12. These correlations are not reported in the table. The variable is percentage of the population living in urban areas, as defined in the *Statistical Abstract,* during the 1960 census.

13. For further evidence on the absence of a relation between contracting and foreign policy votes, see Cobb, "Defense Spending." One other possible reason for the lack of a correlation is the unavoidable aggregation of all kinds of contracting, from exotic aerospace weapons to quasi-civilian goods like food and uniforms. Possibly the former are more potent politically than are the latter.

14. *The Common Defense* (New York: Columbia University Press, 1961), pp. 254-59.

15. "Who Feeds the Military-Industrial Complex? The ABM Fiasco and Congressional Politics," *The Public Life* 1, no. 11 (May 2, 1969): 1-4.

16. The sets were all roll calls in the first and second sessions separately and Republican and Democratic senators separately. The computer program was not able to handle all roll calls and senators for both sessions simultaneously. The finding needs to be further developed with the entire set.

17. Duncan MacRae, *Dimensions of Congressional Voting* (Berkeley and Los Angeles: University of California Press, 1958), and Leroy N. Rieselbach, *The Roots of Isolationism* (Indianapolis: Bobbs-Merrill, 1966).

18. As expressed in chapter 2, I am again deeply indebted to John McCarthy for providing me with this scale. It includes votes on housing and urban development, civil rights generally, open housing legislation, funds for the Office of Economic Opportunity, Headstart, riot insurance, and some "law and order" questions. It also picks up a few defense and aerospace issues. The most conservative end of the spectrum is composed of 19 southern senators, three northern Republicans, and one northern

Democrat. Three other southerners (Baker, Yarborough, and Gore, in increasing order of liberalism) are found in the middle third of all senators.

19. We report the DoD personnel relationship here, as we did not in table 3.5, because this is the first time that its relationship to a voting alignment has been notably higher than that of the military payroll report.

Chapter 4

1. Robert E. Osgood, *Alliances and American Foreign Policy* (Baltimore: The Johns Hopkins Press, 1968), p. 44.

2. Henry Kamm, "Brezhnev Sets the Clock Back," *New York Times Magazine,* Aug. 10, 1969, p. 22.

3. Paul A. Samuelson, "The Pure Theory of Public Expenditure," *Review of Economics and Statistics* 36 (1954): 387-89; "Diagrammatic Exposition of a Theory of Public Expenditure," *Review of Economics and Statistics* 37 (1955): 350-56; and "Aspects of Public Expenditure Theories," *Review of Economics and Statistics* 40 (1958): 332-38; John G. Head, "Public Goods and Public Policy," *Public Finance* 173 (1962): 197-219; Mancur Olson, *The Logic of Collective Action* (Cambridge: Harvard University Press, 1965); Mancur Olson and Richard Zeckhauser, "An Economic Theory of Alliances," in Bruce M. Russett, ed., *Economic Theories of International Politics* (Chicago: Markham, 1968); and "Collective Goods, Comparative Advantage and Alliance Efficiency," in R. N. McKean, ed., *Issues in Defense Economics* (New York: Columbia University Press, 1967). More recently see Jacques M. van Ypersele de Strihou, "Sharing the Defense Burden Among Western Allies," *Yale Economic Essays* 8 (Spring 1968): 261-320, and Philip M. Burgess and James A. Robinson, "Alliances and the Theory of Collective Action: A Simulation of Coalition Processes," *Midwest Journal of Political Science* 13 (May 1969): 194-218.

4. The noncontiguous states of Hawaii and Alaska perhaps constitute minor exceptions in terms of both properties.

5. Pp. 268-69.

6. See Bruce M. Russett, "The Calculus of Deterrence," *Journal of Conflict Resolution* 7 (1963): 97-109. As we shall see below, this is reflected in the relative proportion of the GNP's which Canada and Turkey devote to defense: in 1967, 3.7 percent and 5.6 percent respectively.

7. "An Economic Theory," p. 32.

8. Olson, *The Logic of Collective Action,* pp. 33-34. See also pp. 44-52.

9. Ypersele, "Sharing the Defense Burden," pp. 302-10.

10. Olson, and Olson and Zeckhauser, use the term "sub-optimal," apparently to indicate that the total amount of deterrence purchased by the alliance will be less than would be chosen by the same members if they were under a common government with power to assess and collect contributions. But the amount of *deterrence* bought by the alliance will not be less than the members would buy in *isolation.* Rather, total *expenditures* on deterrence will be less than they would be in the absence of the alliance, the amount of the reduction depending on the degree to which deterrence is really a public good. Since under a *common government* the public good character of deterrence would almost totally apply, it is likely both that

the amount of deterrence then purchased would be more than in alliance and that its cost would be less.

11. This distinction is made by both Ypersele and Richard N. Cooper in their comments following Olson and Zeckhauser, "Collective Goods," on pp. 56-57 and 62. See also Frederic L. Pryor's excellent *Public Expenditures in Communist and Capitalist Nations* (Homewood, Ill.: Irwin, 1969), pp. 88-89. The distinction is not, however, drawn by Olson and Zeckhauser.

12. Considering the alternative possible uses of scientific and technological resources, these benefits are probably exaggerated in most discussions. See the following chapter.

13. See also Malcolm Hoag, "Economic Problems of Alliance," *Journal of Political Economy* 65 (1957): 522-34, and Hoag, "What Interdependence for NATO?" *World Politics* 12, no. 3 (1960): 369-90. Olson and Zeckhauser nevertheless expect the collective goods theory to apply overwhelmingly: "However important the non-collective benefits of alliances may be there can be little doubt that above all, alliances produce collective goods." ("An Economic Theory," p. 28) This viewpoint ignores the private goods inherent in the possible side-payments mentioned at the beginning of the chapter, as well as distracts attention from the private goods provided by military expenditures.

14. "An Economic Theory," pp. 39-41; Ypersele, "Sharing the Defense Burden," pp. 315-20, and Pryor, *Public Expenditures in Communist and Capitalist Nations,* pp. 96-98. The same was true of an earlier draft of this chapter, with NATO data from a different source.

15. Ypersele, "Sharing the Defense Burden Among Western Allies," *Review of Economics and Statistics,* p. 529, suggests that because of this factor the French military effort may be understated by perhaps 1 percent of GNP and the Dutch by .5 percent, as compared with other NATO countries. See that entire article and also Pryor, *Public Expenditures,* pp. 99-107, and Emile Benoit and Harold Lubell, "World Defense Expenditures," *Journal of Peace Research* 2 (1966): 97-113. Defense expenditures should not be taken as perfect indicators of defense obtained, either. A. W. Marshall, "NATO Defense Planning: The Political and Bureaucratic Constraints," in Stephen Enke, ed., *Defense Management* (Englewood Cliffs, N.J.: Prentice-Hall, 1967), argues that European countries probably get less defense capability per dollar than does the United States because of their inability to exploit organizational and production economies of scale.

16. "Sharing the Defense Burden," in *Yale Economic Essays,* p. 316.

17. Data collected for the second edition of the *World Handbook of Political and Social Indicators* by Charles L. Taylor and Michael C. Hudson (New Haven: Yale University Press, forthcoming).

18. Yugoslavia still is formally allied with NATO members Greece and Turkey, but the treaty is universally regarded as dead. Note that this relationship and those we shall find for other alliances are not merely artifacts of a worldwide correlation between D/GNP and GNP. For all 121 countries there is no relationship. An earlier study, Russett et al., *World Handbook of Political and Social Indicators* (New Haven: Yale University Press, 1964), p. 270, found a significant but very slight (8 percent of the variance explained) correlation for 81 countries.

19. Again this fits with the theory according to Olson, *The Logic of Collective Action*, p. 46: "Neither a formal organization nor even an informal organization is indispensable to obtain a collective good."

20. Henry A. Kissinger, *The Troubled Partnership: A Reappraisal of the Atlantic Alliance* (New York: McGraw-Hill, 1965), p. 14. The "decline" of U.S. deterrent credibility is the theme of Gen. Pierre Gallois, *The Balance of Terror: Strategy for the Nuclear Age* (Boston: Houghton Mifflin, 1961).

21. "From Western Europe's standpoint, continued reliance on the American guarantee also seems to be attractive in economic and even political terms, although psychologically it is a good deal more problematic. It is attractive because it provides reliable security at minimum economic cost. Its political cost is also low." Harold van B. Cleveland, *The Atlantic Idea and Its European Rivals* (New York: McGraw-Hill, 1966), p. 20.

22. In no sense should our discussions imply normative acceptance or rejection of the theory of collective goods. If one is concerned with "fairness" and accepts a progressive income tax, it is by no means sure that America pays too much. Ypersele (in *Review of Economics and Statistics*, p. 535) compared actual contributions to those that would result from applying the United Kingdom's tax progressively to NATO efforts in 1963. He concluded that by that criterion the United States and German shares were about "right," French and British military expenditures actually too high, and many of the smaller countries too low.

23. Pryor, *Public Expenditures*, Taylor and Hudson, *World Handbook*, and Institute for Strategic Studies, *The Military Balance, 1968-1969* (London: Institute for Strategic Studies, 1968), p. 55.

24. Pryor, *Public Expenditures*, p. 98, reported the overall positive relationship; while he noted that it was not significant, he did not note that it washed out with the USSR removed. The statistical significance levels are as usual given for a one-tailed test, as the null hypothesis was merely of no relationship—negative relationships were not anticipated.

25. The small differences between 1965 and 1967 should not be taken too seriously because of problems of data comparability, but the differences between both of them (especially as they confirm each other) and the measures for 1962 surely do not stem from data error.

26. This also supports the finding of Pryor (*Public Expenditures*, p. 118) who correlated changes in pact members' D/GNP shares over the 1950-62 period, much as we did for NATO. He found that the correlation between Warsaw members' spending and that of the Soviet Union was *always* statistically significant but was significant for NATO members in but 5 of 12 cases, concluding that Warsaw was the more cohesive alliance.

27. Tiny Albania did escape but appears to be paying the predictable price—a D/GNP ratio of about 10 percent, according to Taylor and Hudson, *World Handbook*.

28. For the texts, see Ruth C. Lawson, ed., *International Regional Organizations, Constitutional Foundations* (New York: Praeger, 1962).

29. *U.S. Overseas Loans and Grants*, Special Report prepared for the House Foreign Affairs Committee (Washington: A.I.D. Statistics and Reports Division, March 1965).

30. On a methodological point, this is because in such a highly skewed distribution not even transforming to logarithms was able to neutralize the statistical effect of the United States.

31. In the failure of GNP and defense shares to correlate below the superpower level Rio is like Warsaw in 1956 and 1962, for which we invoked the explanation of coercion. But the very low mean D/GNP ratio for all the Latin American states shows clearly that they are not being coerced to provide large military forces. Some "coercion" may nevertheless apply to membership decisions, since it would have been politically imprudent for any Latin American government to refuse to join the Rio Pact initially, as it would now be difficult to withdraw.

32. It is widely believed that an explicit commitment is required to make deterrence credible, but this may in fact not be true. See Russett, "The Calculus of Deterrence."

33. Osgood, *Alliances and American Foreign Policy,* p. 2.

34. This distinction is a common one in international politics; it is well stated by Inis Claude, *Power and International Relations* (New York: Random House-Knopf, 1962), pp. 115-17.

35. For some thoughts on this, see Charles Wolf, Jr., "Some Aspects of the 'Value' of Less-Developed Countries to the United States," in Russett, *Economic Theories of International Politics.*

Chapter 5

1. In addition to the state-by-state studies cited in chapter 3, Murray L. Weidenbaum, "Problems of Adjustment for Defense Industries," in Emile Benoit and Kenneth E. Boulding, eds., *Disarmament and the Economy* (New York: Harper & Row, 1963), presents data on the distribution of contracts by industry. See also James L. Clayton, "Defense Spending: Key to California's Growth," *Western Political Quarterly* 15 (June 1962): 280-93. *Science* 161 (1968): 448 reports the receipts of various universities for Department of Defense-sponsored research in the physical, biological, and social sciences. On Canada, see Gideon Rosenbluth, *The Canadian Economy and Disarmament* (New York: St. Martin's Press, 1967), and for Britain, Economist Intelligence Unit, *The Economic Effects of Disarmament* (London: Economist Intelligence Unit, 1963).

2. See Milton Friedman, "Why Not a Volunteer Army?" *New Industrialist Review* 4, no. 4 (Spring 1967): 9; Walter Y. Oi, "The Economic Cost of the Draft," *American Economic Review* 57, No. 2 (May 1967): 48-49; and Mark V. Pauly and Thomas D. Willett, "Who Bears the Burden of National Defense?" in James C. Miller, III, *Why the Draft?* (Baltimore: Penguin, 1968), pp. 53-57.

3. Cabinet Co-ordinating Committee on Economic Planning for the End of Vietnam Hostilities, "Report to the President," in *Economic Report of the President* (Washington D.C.: U.S. Government Printing Office, 1969), p. 206. A. C. Fisher, "The Cost of the Draft and the Cost of Ending the Draft," *American Economic Review* 59, no. 3 (June 1969): 239-54, sets $5-7.5 billion dollars as the upper limit.

4. "Shifting the Composition of Government Spending: The Impli-

cations for the Regional Distribution of Income," *Peace Research Society Papers* 5 (1967): 15-43. All the material by Weidenbaum cited here dates from a period preceding his government position and also precedes our 1967-68 data, which showed an improvement in the South's status.

5. For a good review of opportunity costs of defense, see Weidenbaum's "Military Programs and the American Economy," in a forthcoming book being edited by Adam Yarmolinsky for the Twentieth Century Fund. One not widely appreciated cost is the distortion of economic growth that may be forced by high taxes. As former Assistant Secretary of Defense Charles J. Hitch puts it, if military programs were half their size and tax rates proportionately lower, "We wouldn't have to consider tax implications the most important aspect of every business decision." See his "The Military Budget and Its Impact on the Economy," in Thomas K. Hitch, ed., *Economics for the 1960's* (Honolulu: First National Bank of Hawaii, 1961), p. 50.

6. Much of the war's cost is undoubtedly absorbed by other military categories, such as the number of troops kept in other parts of the world and reductions in new procurement and maintenance of old equipment not being used in the war zone. Even so, it is likely that our method understates the cost of the war to the civilian sectors of the system. President Johnson's State of the Union Message in January 1968 put the annual cost at $25 billion in 1967. For 1968 it was nearly $29 billion.

7. *The Weapons Culture* (New York: Norton, 1968), p. 173.

8. "Adjusting to a Defense Cutback: Public Policy Toward Business," *Quarterly Review of Economics and Business* (Spring 1964), pp. 7-14, cited by Weidenbaum, "Defense Expenditures and the Domestic Economy," in Stephen Enke, ed., *Defense Management* (Englewood Cliffs, N.J.: Prentice-Hall, 1967), p. 329.

9. U.S. Senate, Committee on Labor and Public Welfare, Subcommittee on Employment and Manpower, *Hearings: Nation's Manpower Revolution, Part 9* (Washington: U.S. Government Printing Office, 1964), p. 3146.

10. William L. Baldwin, *The Structure of the Defense Market, 1955-1964* (Durham, N.C.: Duke University Press, 1967), p. 146. See also Seymour Melman, *Our Depleted Society* (New York: Holt, Rinehart, and Winston, 1965).

11. This is not the only plausible way to process these data, though I believe it to be at least as appropriate as any other. One problem involves the choice of years, since much of the variation discussed here stems from differences between the World War II years and the following decades rather than within these periods. Excluding 1939-45, however, would sharply reduce both the number of data points and the range of experience being analyzed. I have been careful to examine the scattergram for each relationship with this in mind. For a detailed examination of this and other methodological problems, and alternative modes of analysis and their consequences, see Bruce M. Russett, "Some Decisions in the Regression Analysis of Time-Series Data," in Joseph L. Bernd, ed., *Mathematical Applications in Political Science* V (Charlottesville: University Press of Virginia, 1970). This article also has some important substantive conclusions on the difference between wartime and peacetime defense efforts, but generally it

is clear that methodological considerations do not notably affect the findings. The same article shows that objections to correlating ratios do not apply here. All the data analyzed in this and the following chapter are available on cards from the World Data Analysis Program, Yale University, 89 Trumbull Street, New Haven, Connecticut 06520.

12. This index is computed by the formula

$$IPR = \frac{25 \text{ (reg. coef.)}}{400 \text{ (mean prop. of GNP)}}$$

for the dependent variable. The choice of illustrative values for the GNP and defense increases does not of course imply that the impact is thus in any particular buildup.

13. "The Garrison State," *American Journal of Sociology* 46, no. 4 (January 1941): 455-69.

14. P. V. Sokolov, ed., *Military-Economic Problems in a Political Economy Course* (Moscow: Voyenizdat, 1968; translated and published, Washington: U.S. Joint Publications Research Service, 1969), pp. 72, 19.

15. The high percentage of variance explained is nonetheless a bit deceptive, as a close examination of the plot discloses. If one looks at the periods of moderate defense expenditure one finds only a mild relationship between the two variables. Most of the variance is concentrated on the differences between the moderate and high defense groups and within the latter.

16. Robert M. Solow, *Capital Theory and the Rate of Return* (Amsterdam: North Holland, 1963).

17. The costs of military procurement abroad can of course be figured more precisely and directly than we have done here, but by missing the indirect effects of inflation and diverted demand such a computation would understate the loss. Defense Department calculations of the direct cost do in fact come to but $1.5 billion; a complex independent analysis that includes the indirect effects suggests $4 billion, or that without the war the United States would actually have maintained a balance of payments surplus. See Leonard Dudley and Peter Passell, "The War in Vietnam and the United States Balance of Payments," *Review of Economics and Statistics* 50, no. 4 (November 1968): 437-42. During 1960-67 the United States ran an annual average international balance, on a liquidity basis, of −$2.5 billion. The balance on private account, however, was +$.3 billion, with a −$2.8 billion loss in government transactions. Hence the entire recent balance of payments deficit can be attributed to government activities, primarily military expenditures and, to a lesser degree, foreign aid. See "Pax Americana and the U.S. Balance of Payments," *New England Economic Review,* January/February 1969, p. 43.

18. Although recent experience may make it seem obvious that public civil expenditures are likely to be inversely related to defense shares, this perception has not always been universal. Several years ago W. Glenn Campbell reported a strong inverse relationship for the 1953-63 period, in "Assuring the Primacy of National Security," in David M. Abshire and Richard V. Allen, *National Security: Political, Military and Economic Strategies in the Decade Ahead* (New York: Praeger, 1963), pp. 963-84.

Otto Eckstein, however, noted some questionable assumptions in Campbell's analysis and followed with a stronger expression of doubt: "I think that historical experience has been that governments are either stingy, or they're spenders. And if they're stingy about defense, they're stingy about everything. I would say that the historical record suggests that the association between civilian spending and military spending is positive, not negative." "Discussion" in ibid., p. 1012. Samuel P. Huntington, *The Common Defense* (New York: Columbia University Press, 1961), p. 22, says that during the Truman and Eisenhower administrations (except for the Korean War period) domestic civil claims had first priority, with defense forced to accept whatever resources were left over. Weidenbaum ("Military Programs and the American Economy") maintains that the situation is now exactly reversed: defense has first claim, and civilian needs must settle for the residual.

19. *The Common Defense,* p. 208. Further masking of the impact on actual programs may stem from the inability of government agencies to reduce costs for building maintenance and tenured employees, thus forcing them in dry times to cut other expenses disproportionately.

20. It might be thought that this is a case where, if there were no immediate effects of defense needs on state and local government finances, defense might nevertheless force some delayed cutbacks by the areal units. But this too is not the case, since an effort to lag local expenditures a year or two behind defense made no improvement in the fit. Apparently the federal and the areal units of government are sufficiently independent in their major revenue sources that fluctuations in the needs of the former do not seriously hamper the latter.

21. Quoted in Julius Duscha, *Arms, Money, and Politics* (New York: Ives Washburn, 1965), p. 18.

22. Edward F. Denison, *Sources of Economic Growth in the United States and the Alternatives Before Us* (C.E.D., 1962) as cited in Ruth P. Mack, "Ecological Processes in Economic Change," *American Economic Review* 58, no. 2 (May 1968): 47. See also Gary S. Becker, *Human Capital: A Theoretical and Empirical Analysis* (New York: National Bureau of Economic Research and Columbia University Press, 1964), and Theodore W. Schultz, "Investment in Human Capital," *American Economic Review* 51, no. 1 (March 1961): 1-17. On this idea of educated manpower as a form of national capital see also the chapter by Kenneth Boulding in Walter Adams, ed., *The Brain Drain* (New York: Macmillan, 1968).

23. U.S. Bureau of the Census, *Statistical Abstract of the United States, 1968* (Washington: U.S. Government Printing Office, 1968), p. 526, and *Statistical Abstract of the United States, 1963* (Washington: U.S. Government Printing Office, 1963), p. 544. Comparable data for earlier years are not available, so this variable was not used in the computations for table 5.2.

24. Professor Herman Somers of Princeton University has pointed out, in a personal communication, that much the same has happened in the private sector. The real GNP shares of both personal health expenditures and total health (public and private) declined over the peri-

od. This fact, not generally known, should be a major cause of concern.

25. Myron Brenton, "The Higher Cost of Higher Education," *New York Times Magazine,* April 28, 1968, p. 32.

Chapter 6

1. *Competition and Complementarity Between Defense and Development—A Preliminary Approach,* P-1743 (Santa Monica, Calif: RAND Corporation, 1959).

2. *Public Expenditures in Communist and Capitalist Nations* (Homewood, Ill.: Irwin, 1969) p. 124, also 120-21.

3. Data collected under the direction of Charles L. Taylor and Michael C. Hudson, World Data Analysis Program, Yale University. Skewed distributions were subjected to logarithmic transformations before correlating.

4. Bruce M. Russett et al., *World Handbook of Political and Social Indicators* (New Haven: Yale University Press, 1964), pp. 269-70.

5. On these matters see William P. Snyder, *The Politics of British Defense Policy, 1945-1962* (Columbus: Ohio State University Press, 1964), and Harold and Margaret Sprout, "The Dilemma of Rising Demands and Insufficient Resources," *World Politics* 20, no. 4 (July 1968): 660-93.

6. Note the similar conclusion of Sprout and Sprout, "The Dilemma of Rising Demands." Samuel P. Huntington, *The Common Defense* (New York: Columbia University Press, 1961), p. 243, reports that in a nationwide survey asking Britishers: "If the government wants to cut down its spending, which of these would you put first?" the defense option was chosen by 32 percent. Similarly, in France, Norway, and Poland a proportion ranging from nearly half to over two-thirds of those with opinions preferred to cut defense spending rather than leave it at current levels or raise it. See Johan Galtung, "Public Opinion on the Economic Effects of Disarmament," in Emile Benoit, ed., *Disarmament and World Economic Interdependence* (Oslo: Universitetsforlaget, 1967), p. 173. This is a marked contrast to the standard situation in the United States, as noted on page 23 above.

7. Peacock and Wiseman, *Growth of Public Expenditures,* pp. 24-30.

Chapter 7

1. The basic studies remain Samuel P. Huntington, *The Soldier and the State* (Cambridge: Harvard University Press, 1959), and Morris Janowitz, *The Professional Soldier* (Glencoe, Ill.: Free Press, 1960).

2. *Political Order in Changing Societies* (New Haven: Yale University Press, 1968), chap. 2.

3. Whites currently outnumber blacks by about eight to one and have an average income level a little over half again as high. It is worth noting that the United States ranks as a little more unequal among world societies when the measure employed is an overall (Gini) index that includes the conditions of the poorest sectors than when one looks only at the proportion of income going to the richest 10 percent of the population.

On the latter, see Bruce M. Russett et al., *World Handbook of Political and Social Indicators* (New Haven: Yale University Press, 1964), pp. 243-47.

Appendix

1. Richard Brody and John Vesecky, "Soviet Openness to Changing Situations: A Critical Evaluation of Certain Hypotheses About Soviet Foreign Policy Behavior" (Stanford, Stanford University, 1965).

2. A partial exception, examined in this review, is Jeffrey S. Milstein and William C. Mitchell, "Computer Simulation of International Processes: The Vietnam War and the Pre-World War I Naval Race," *Peace Research Society Papers* 12 (1970). However, such empirical results are much less likely in studies of the more rapidly changing contemporary world.

3. As should be clear from the body of this paper, we are not demanding that all terms must be directly measurable. The point is that every term must be meaningful.

4. Richardson's major effort is contained in: Lewis F. Richardson, *Arms and Insecurity: A Mathematical Study of the Causes and Origins of War* (Pittsburgh and Chicago: Boxwood and Quadrangle, 1960).

5. P. E. Chase, "The Relevance of Arms Race Theory to Arms Control," *General Systems* 13 (1968): 91-98. See p. 91.

6. Arthur Lee Burns, "A Graphical Approach to Some Problems of the Arms Race," *Journal of Conflict Resolution* 3 (1959): 326-42; Martin C. McGuire, *Secrecy and the Arms Race* (Cambridge, Mass.: Harvard University Press, 1965).

7. In addition to Richardson's *Arms and Insecurity,* a succinct presentation of his model can be found in Lewis F. Richardson, "Mathematics of War and Foreign Policies" in James R. Newman, ed., *The World of Mathematics* (New York: Simon and Schuster, 1956), 2:1240-53.

Summaries and criticisms of Richardson's work can be found in the following sources: William R. Caspary, "Richardson's Model of Arms Races: Description, Critique, and an Alternative Model," *International Studies Quarterly* 11, no. 1. (March 1967): 63-90; McGuire, pp. 33-38; Anatol Rapoport, "Lewis F. Richardson's Mathematical Theory of War," *Journal of Conflict Resolution* 1 (1957): 249-99; Anatol Rapoport, *Fights, Games and Debates* (Ann Arbor: University of Michigan Press, 1960); Thomas L. Saaty, *Mathematical Models of Arms Control and Disarmament* (New York: Wiley, 1968).

8. Saaty, p. 46.

9. The equations and definitions of terms are presented in Richardson, *Arms and Insecurity,* pp. 14-16.

10. An interesting discussion of Richardson's "determinism" is found in Rapoport, pp. 249-52. This review's discussion of the matter only seems to be at odds with Rapoport's. Our argument is that Richardson is "deterministic" only in the sense that the concept of probability is not explicit in his system.

11. Richardson, *Arms and Insecurity,* p. 12ff.

12. Ibid., pp. 32-33; Richardson, "Mathematics of War and Foreign

Policies," p. 1246. Richardson uses a somewhat different notation.

13. Richardson, *Arms and Insecurity,* pp. 32-34.

14. This example is taken from ibid., pp. 20-21.

15. Ibid., p. 20.

16. Rapoport, *Fights, Games and Debates,* pp. 43-44.

17. A simple, clear explanation of "equilibrium" and "stability" is in William J. Baumol, *Economic Dynamics* (2d ed. New York: Macmillan, 1956). See pp. 213-17 for definitions in terms of difference equations. The same concepts are explained in the context of differential equations on pp. 301-03.

It might be noted that, for the sake of simplicity, this paper has ignored the possibility of "moving equilibria" (i.e. particular solutions which are functions of time). This is because, not knowing the parametric values or the initial conditions, it is not possible to determine whether the particular solution is a constant or a function of time. In substantive terms, the presence of a moving rather than a constant equilibrium would mean that, instead of a point where neither side increased its force levels, there would be a situation where each side increased its strength by some proportion during each time period.

18. Caspary, "Richardson's Model of Arms Races."

19. Ibid., p. 67.

20. Ibid., p. 69; Richardson, *Arms and Insecurity,* p. 35.

21. *Arms and Insecurity* contains a great deal more material than is described in this review. For instance, Richardson offers an "n" nation model. However, his more complex formulations seem to have even more problems than the model reviewed here.

22. The following discussion is based on Paul Smoker, "The Arms Race: A Wave Model," *Peace Research Society (International) Papers* 4 (1966): 151-92, and Smoker, "The Arms Race as an Open and Closed System," *Peace Research Society (International) Papers* 7 (1967): 41-62. A critique of the second article which is also applicable to the first is J. G. Dash, "Comments on the Paper by Smoker," *Peace Research Society (International) Papers* 7 (1967): 63-65.

23. Smoker, "The Arms Race as an Open and Closed System," p. 54.

24. A clear explanation of harmonic motion is found in Robert Becker, *Introduction to Theoretical Mechanics* (New York: McGraw-Hill, 1954). Pages 136-37 are particularly relevant.

25. Ibid., p. 55, equation J.

26. For substantive definitions of terms, see Smoker, "The Arms Race as an Open and Closed System," p. 55.

27. Smoker, "The Arms Race: A Wave Model," pp. 154-58.

28. As Jeffrey Milstein suggested in a personal communication, the first-order derivative (or difference) of spending with respect to time might well oscillate about zero.

29. Caspary, "Richardson's Model of Arms Races"; we have followed Caspary's format closely.

30. Ibid., pp. 69-70. We have altered Caspary's notation to correspond with Richardson's.

31. This argument and the notational change used in equations 18 and

19 taken from William R. Caspary, "Formal Theories of Reaction Processes in International Relations" (paper presented at the American Political Science Association Annual Meeting, September 1969), pp. 35-38. The argument concerning Richardson in this article is basically the same as that in his "Richardson's Model of Arms Races." The rest of "Formal Theories of Reaction Processes" is quite different and should be of great interest when published in final form.

32. This is Caspary's "model 2" in "Richardson's Model of Arms Races," pp. 72-77.

33. The role of qualitative change and the general issue of qualitative arms races are discussed in Samuel P. Huntington, "Arms Races: Prerequisites and Results," *Public Policy* 8 (1958): 41. Although Huntington's article is beyond the purview of this essay, it is one of the best analyses in this field. The importance of normal obsolescence was suggested to me by Gary Brewer.

34. Caspary demonstrates this point in "Richardson's Model of Arms Races," p. 74, note 12. To expand on this a bit, we may note that for small x,

$$p\frac{dx}{dt} = a(C - Mx)(1 - e^{-ND/C}) \approx aC(1 - e^{-ND/C})$$

Expanding $e^{-ND/C}$ by the Taylor series:

$$e^{-ND/C} = 1 - ND/C + \frac{1}{2!}(\frac{ND}{C})^2 - \frac{1}{3!}(\frac{ND}{C})^3 \ldots$$

When ND/C is small (i.e. when the desired increase in spending is only a small fraction of the total amount that could be spent on armaments), then:

$$e^{-ND/C} \approx 1 - ND/C$$

Substituting into equation 20, we get

$$p\frac{dx}{dt} \approx aN(\frac{ky}{a} - x + \frac{g}{a})$$

Since p and N are dimensional constants included only to convert from force units to money units, they may be neglected (actually Richardson seems to measure x and y directly in money units).

35. This means that $f(D) = 1 - e^{-ND/C}$ grows more slowly than D. This can be written

$$\frac{d}{dD}f(D) < 1.$$

But

$$\frac{d}{dD}f(D) = \frac{N}{C}e^{-ND/C}$$

As the exponent grows the term very rapidly becomes less than unity.

36. As ND becomes very large, equation 20 reduces to

$$p\frac{dx}{dt} = p(C - Mx).$$

This follows from the observation that $e^{-ND/C} \to 0$ as $ND \to \infty$ Furthermore, where ND is very large, Mx is small compared to C because ND + Mx = C. Hence expenditures asymptotically approach C.

37. For the following demonstration see Caspary, "Richardson's Model of Arms Races," pp. 76-77.

38. Ibid., p. 77. See also chapter 1 of this book.

39. This is model 3 in ibid., pp. 77-86.

40. Murray Wolfson, "A Mathematical Model of the Cold War," *Peace Research Society (International) Papers* 9 (1968): 107-23.

41. Ibid., p. 111.

42. Wolfson, p. 112. This is Wolfson's Model I.

43. Wolfson recognizes this problem. See ibid., pp. 112-13.

44. Bernard Fall, *The Two Vietnams* (rev. ed. New York: Praeger, 1964); see especially chapter 2. Ho had some basis for believing that a new status quo had been established, or was at least imminent, as indicated by Fall's observation that had the planned elections been held in July 1956 Ho Chi Minh would have gained control of South Vietnam. See p. 233.

45. The problem of assuming the parameters to be constant is dealt with in Wolfson, p. 112.

46. The following demonstration is based on Wolfson, pp. 112, 114 (note 8), 115-16. These principles are explained in Baumol, *Economic Dynamics,* pp. 362-65.

47. Baumol, p. 362.

48. The use of inequalities 26 and 27 diverges from Wolfson's analysis because of an error in the latter. I am grateful to Dr. Augustine Tan for pointing this out to me. However, Wolfson's mistake does not invalidate his conclusion that the coefficients must be tightly constrained if the system is to be stable.

In particular, from (a) $-2 < a_s - a_u + \kappa_s + \kappa_u < 2$, Wolfson deduces (b) $|\frac{a_s - a_u}{2}| < 1 - \frac{\kappa_s + \kappa_u}{2}$. See Wolfson, p. 115. However (b) implies that (c) $-2 + \kappa_s + \kappa_u < a_s - a_u < 2 - \kappa_s - \kappa_u$. But $-2 + \kappa_s + \kappa_u < a_s - a_u$ implies (d) $-2 < a_s - a_u - \kappa_s - \kappa_u$ and $a_s - a_u < 2 - \kappa_s - \kappa_u$ implies (e) $a_s - a_u + \kappa_s + \kappa_u < 2$. Clearly inequalities (d) and (e) are incompatible with (a).

However, let us write (e) in the equivalent form:

(f) $\frac{a_s - a_u}{2} < 1 - \frac{\kappa_s + \kappa_u}{2}$

Then, if $a_s > a_u$, $2 > \kappa_s + \kappa_u$

$$\text{if } a_s = a_u, 0 < 1 - \frac{\kappa_s + \kappa_u}{2}$$

$$\text{or } 1 > \frac{\kappa_s + \kappa_u}{2}$$

$$\text{or } 2 > \kappa_s + \kappa_u$$

$$\text{if } a_s < a_u, 0 < 1 - \frac{\kappa_s + \kappa_u}{2}$$

$$\text{or } 2 > \kappa_s + \kappa_u$$

Since it can be shown that inequality (d) places no restriction on the sum of $\kappa_s + \kappa_u$, it follows that $\kappa_s + \kappa_u < 2$. Finally, let us assume each of these coefficients to be positive. Then, returning to the text, 26a implies

$\dfrac{a_s - a_u}{2} < 1$ and 26b requires that $-2 < \dfrac{a_s - a_u}{2}$.

49. Wolfson, p. 113.

50. Wolfson deals with some difficulties in two more complex systems, deducing some rather interesting new results. See ibid., pp. 116ff.

51. Ibid., p. 114 (note 8), pp. 115-16, conditions 2 and 3; also see Baumol, pp. 362-64, propositions 8 and 9.

52. Wolfson, p. 113. Wolfson assumes that a_s and a_u are greater than unity and are "probably rather large." On the other hand, β_s and β_u are assumed to be less than unity.

53. Milstein and Mitchell, "Computer Simulations of International Processes." Only the first section of their article is reviewed here, since the section on Vietnam is not germane to arms race matters.

54. Richard P. Lagerstrom and Robert C. North, "An Anticipated Gap, Mathematical Model of International Dynamics" (unpublished, Stanford University, April 1969). I have not reviewed this paper because it was available only in rough draft.

55. The solution to equations 30 and 31 is of the form:

$$x = C_1 a_1 e^{\lambda_1 t} + C_2 a_2 e^{\lambda_2 t} + C_3 a_3 e^{\lambda_3 t} + C_4 a_4 e^{\lambda_4 t}$$

$$+ Z_1 (t)$$

$$y = C_1 \beta_1 e^{\lambda_1 t} + C_2 \beta_2 e^{\lambda_2 t} + C_3 \beta_3 e^{\lambda_3 t} + C_4 \beta_4 e^{\lambda_4 t}$$

$$+ Z_2 (t)$$

The meaning of this may be found in any text on differential equations. Wilfred Kaplan, *Elements of Differential Equations* (Reading, Mass: Addison-Wesley, 1964) gives a succinct description. See especially chapter 7.

Let it suffice here to note that the values of $\lambda_{1.....4}$ obviously affect the speed with which equilibrium is approached. The four values of λ are derived from the following "characteristic equation":

$$\lambda^4 + (d + c) \lambda^3 + (a + b + d - m) \lambda^2 + (ad - cb - lm - kn) \lambda$$
$$+ ab - kl = 0$$

Thus it is clear that all the parameters affect λ.

56. Milstein and Mitchell, p. 122. We have used the regression equations based on cumulative rather than regular data since the authors report that the former type provides better predictions.

57. Ibid., pp. 122-24. Kendall Moll, *The Influence of History Upon Seapower, 1865-1914* (Stanford: Stanford Research Institute, 1968), uses, on the basis of some historical information, different desired-strength ratios for different prime ministers.

58. The use of Routh's criterion is suggested in Lagerstrom and North,

p. 52. However, a more complete explanation of this theorem is presented in Baumol, *Economic Dynamics,* p. 303. The substitution of the parametric values of equations 34 and 35 into Baumol's formulation shows that the system should be stable. It might be noted, however, that the criterion would not be fulfilled if these parameters were slightly different since the value of the 3 x 3 determinant, in Baumol's presentation, is only 0.01.

59. See McGuire, *Secrecy and the Arms Race,* p. 88, for the argument that various tradeoffs are possible between the minimum *number* of missiles one wants to survive and the size of the *probability* that the goal will be realized. Indeed, McGuire's analysis contains many examples of the necessity to balance competing needs. It might be noted that this study shows the meaninglessness of the belief that a government can simply spend what is "necessary" to meet its security requirements without striking compromises—both among strategic goals and between military and other needs. Also relevant to these issues is Charles J. Hitch and Roland N. McKean, *The Economics of Defense in the Nuclear Age* (Cambridge, Mass: Harvard University Press, 1960).

60. For example, Burns, "A Graphical Approach to Some Problems of the Arms Race;" Michael D. Intriligator, "Some Simple Models of Arms Races," *General Systems* 9 (1964): 143-47; Eric Moberg, "Models of International Conflicts and Arms Races," *Cooperation and Conflict, Nordic Studies in International Politics* 2 (1966): 80-93. Game theoretic analyses have also been used very effectively in the study of strategy and arms race processes. To cite just two works: Thomas C. Schelling, *The Strategy of Conflict* (Cambridge, Mass: Harvard University Press, 1960); and Rapoport, *Fights, Games, and Debates.*

61. McGuire, pp. 83-84. It is this "second approximation" upon which the interaction model is built.

62. See ibid., p. 143, for equations 36 and 37. The terms are defined on pp. 87-88 and on the "notation" chart facing p. 1. Although the preliminary model embodied in equations 36 and 37 does not include cost constraints, these are introduced in McGuire's more complex formulations. However, they cannot be covered in this review.

63. Ibid. One of his assumptions is that an attacking missile can destroy no more than one of the defender's missiles. There is, however, the possibility that more than one of the defender's missiles, or launching facilities controlling several second-strike missiles, might be made inoperative by a single attacking missile. Along these lines, the problems introduced by MIRV systems are serious.

64. Ibid., chapters 4, 6, and 8.

65. Caspary and Milstein are now attempting to construct simulation models.

Index

DATE DUE